THE LIGHTER SIDE
OF GRAVITY

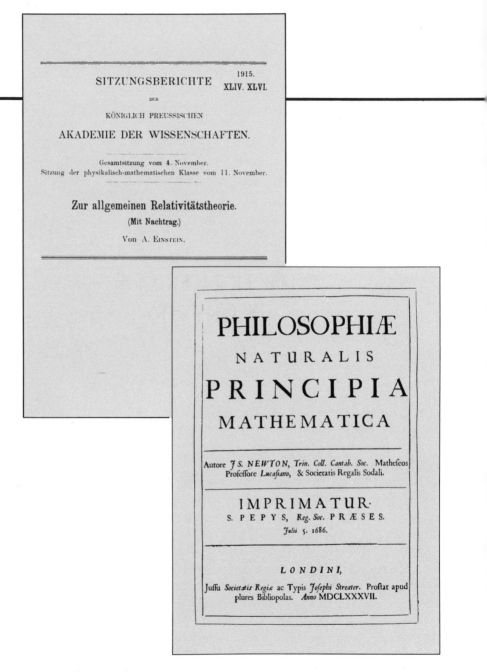

These are the cover pages of Newton's 1686 *Principia* and Einstein's 1915 paper on general relativity. These works gave us the foundations for our two greatest theoretical advances in understanding gravity. (*Principia* page through the courtesy of the Bancroft Library, University of California, Berkeley.)

THE LIGHTER SIDE
OF GRAVITY

JAYANT V. NARLIKAR

Tata Institute of Fundamental Research
Bombay, India

W. H. Freeman and Company
San Francisco

Project Editor: Larry Olsen
Designer: Robert Ishi
Production Coordinator: Bill Murdock
Illustration Coordinator: Cheryl Nufer
Artist: Dale Johnson
Compositor: Composition, etc.
Printer and Binder: The Maple-Vail
                            Book Manufacturing Group

Library of Congress Cataloging in Publication Data

Narlikar, Jayant Vishnu, 1938–
    The lighter side of gravity.

    Includes index.
    1. Gravity. I. Title.
QB331.N37   521'.1   81-19496
ISBN 0-7167-1343-8   AACR2
ISBN 0-7167-1344-6 (pbk.)

Printed in the United States of America

9 8 7 6 5 4 3 2 1 MP 0 8 9 8 7 6 5 4 3 2

Contents

Preface

It is often said that modern theoretical physics began with Newton's law of gravitation. There is a good measure of truth in this remark, especially when we take into account the aims and methods of modern physics—to describe and explain the diverse and complex phenomena of nature in terms of a few basic laws.

Gravity is a basic force of the Universe. From the motions of ocean tides to the expansion of the Universe, a wide range of astronomical phenomena are controlled by gravity. Newton summed up gravity in his simple inverse-square law. Einstein saw in it something of deeper significance that linked it to space and time. The modern theoretical physicist is trying to accommodate it within a unified theory of all basic forces. Yet, gravity remains an enigma today.

In this book I have attempted to describe the diversity, pervasiveness, and importance of this enigmatic force. It is fitting that I have focused on astronomical phenomena because astronomy is the subject that first provided and continues to provide a testing ground for the study of gravity. These phenomena include the motions of planets, comets, and satellites; the structure and evolution of stars; tidal effects on the Earth and in binary star systems; highly dense objects, such as neutron stars, black holes, and white holes; and the origin and evolution of the Universe itself.

The presentation throughout the book is at a nontechnical level. Although the title is *The Lighter Side of Gravity*, it should not be mistaken for a nonserious presentation. My aim throughout this book has been to seriously emphasize the mutually beneficial interaction between astronomy and gravitational theory, an interaction that has lasted for three centuries and is bound to last into the foreseeable future.

The desire to keep the book compact and more or less along conventional lines has meant the omission of many interesting and fruitful ideas about gravity outside the frameworks of Newton and Einstein. Contrary to the view held by the conservative majority of astronomers, I feel that the last word on gravity has not yet been said and that some astronomical phenomena already warrant a fresh input of ideas. The emphasis in this book is, however, on the successes of the ideas of Newton and Einstein and on the daring and speculative applications that these ideas have inspired.

I have enjoyed writing this book, a job that was made easier because of help from so many. Initial comments by John Faulkner were most helpful. My wife Mangala made the initial sketches of the nontechnical figures in the book. She and my parents read the first draft and made valuable suggestions to make it more readable to the lay person. It is a pleasure to acknowledge the prompt assistance of the staff of the drawing section and the photography section of T.I.F.R. in the preparation of some of the technical figures. Artwork from other sources has been duly acknowledged at the appropriate places. Finally, my thanks are due to Mr. D. B. Sawant for typing the manuscript expeditiously.

Tata Institute of Fundamental Research   *Jayant V. Narlikar*
Bombay, India

THE LIGHTER SIDE
OF GRAVITY

A time-lapse photograph of the night sky shows the position
of the Pole Star and the apparent motion of the stars.
(Courtesy of Lick Observatory.)

1

Why Things Move

1

Why Things Move

The restless Universe

From ancient Hindu mythology comes this story about the Pole Star: King Uttanapada had two wives. The favorite, Suruchi, was haughty and proud, while the neglected Suniti was gentle and modest. One day Suniti's son Dhruva saw his step-brother Uttama playing on their father's lap. Dhruva also wanted to join him there but was summarily repulsed by Suruchi, who happened to come by. Feeling insulted, the five-year-old Dhruva went in search of a place from where he would not have to move. The wise sages advised him to propitiate god Vishnu, which Dhruva proceeded to do with a long penance. Finally Vishnu appeared and offered a boon. When Dhruva expressed his wish, Vishnu placed him in the location now known as the Pole Star—a position forever fixed.

Unlike other stars and planets, the Pole Star does not rise and set; it is always seen in the same part of the sky. This immovability of the Pole Star has proved to be a useful navigational aid to mariners from ancient to modern times. Yet, a modern-day Dhruva could not be satisfied with the Pole Star as the ultimate position of rest. Let us try to find out why.

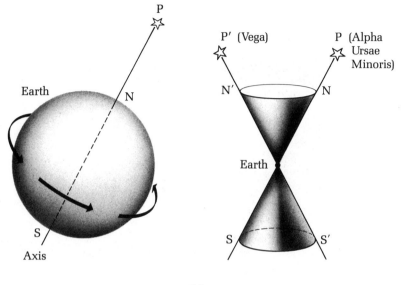

1-1. (a) The Earth rotates about its north–south axis. The Pole Star
P at present lies almost along this axis. (b) The north–south
axis is not fixed in space; it *precesses*, or traces the shape of a
cone on the sky. Hence, relative to the axis, the Pole Star
appears to change direction. Two extreme positions of the
axis NS and N'S' are shown as it traces the shape of a cone.
At present, the north–south axis points toward the star
Alpha Ursae Minoris; after 13,000 years, it will point toward
the star Vega.

The Pole Star does not appear to change its direction in the sky
because it happens to lie more or less along the Earth's axis of
rotation. As the Earth rotates about its axis, other stars rise over the
eastern horizon and set over the western horizon. But as long as
the Earth's rotation axis remains unchanged in its direction, the
Pole Star will not rise and set but instead will appear fixed, staying
always in the same direction. The Earth's rotation axis does,
however, change its direction very slowly. Instead of being fixed as
in Figure 1-1a, it describes a narrow cone as shown in Figure 1-1b.
The time taken for one revolution of the axis along this cone is
nearly 26,000 years. No wonder then that, over the human lifetime,

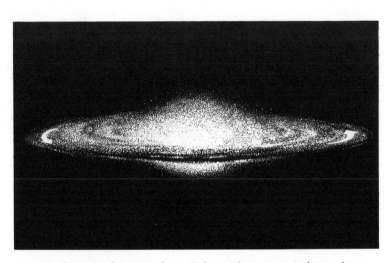

1-2. A schematic diagram of our Galaxy. The arrows indicate the direction the Galaxy rotates. The approximate location of the Pole Star is shown by a circled cross.

or indeed over several centuries, the Pole Star hardly appears to move, whereas in fact it is slowly changing its direction with respect to the Earth's axis.

But this is not the real problem! The Pole Star itself is not fixed in space. Like other stars in our Galaxy, it is moving. Indeed, the Galaxy as a whole (which, as shown in Figure 1-2, is a disk-shaped object with a small bulge in the middle and contains some 100 billion stars) rotates about *its* axis with a period of nearly *200 million* years.* So Dhruva cannot really claim to have found a fixed, immovable place!

Indeed, a closer examination shows that mobility rather than rest is the characteristic feature of the Universe. Just as the astronomer is discovering examples of motion on the large scale, so is the student of microscopic physics finding various examples of motion on the small scale. When we look at a river from a distance,

---

*Apart from this rotation, our Galaxy takes part in a large-scale motion generally known as the expansion of the Universe. But more of this in Chapter 10.

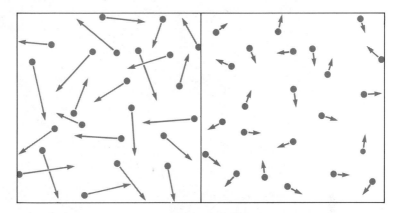

1-3. Air molecules in random motion. The length of the arrow indicates the speed of the corresponding molecule. In the distribution on the right, the speeds are lower, corresponding to a lower air temperature than in the distribution shown at left.

it may appear to us to be at rest. However, when we approach it, we begin to see the steady flow of water. On a windless day, we might imagine the air to be at rest. Microscopic physics will tell us, however, that the still air is made up of molecules in *random motion*, and this random motion endows the air with the property of *temperature* (Figure 1-3). The faster this random motion, the higher the air temperature. Going down to the atomic level, we continue to find motion in some form or other in all types of matter. There are lattice vibrations in crystals; there are negatively charged electrons moving freely inside metals; there are bound electrons jumping from one orbit to another inside atoms. Even inside atomic nuclei, things are not at rest! In high-energy particle accelerators like the Fermilab near Chicago or CERN near Geneva, atomic physicists are gradually discovering the secrets of the strange world of elementary particles (Figure 1-4). In such a restless universe, it is going to be a futile exercise to look for rest and immovability. Rather, we should ask the question *"Why do things move?"*

1-4. A section of the high-energy particle accelerator, the Fermilab. (Courtesy of R. R. Wilson, National Accelerator Laboratory.)

From Aristotle to Galileo

This question was posed some twenty-three centuries ago by Aristotle, a Greek philosopher, a pupil of Plato, and the tutor of Alexander the Great. Aristotle's philosophy dominated Western thought through the Middle Ages, and his science had the authority of the Roman Catholic Church behind it. Today, living in the age of science, we find Aristotle's approach and ideas strange and difficult to grasp. Yet, when seen against the background of Greece of 350 BC, they reflect a highly systematic attitude.

Aristotle talked about *change* in a system in general terms, and to him motion meant *local* change. Everyday observations present several examples—the motion of stars across the sky, the rising of smoke, the motion of clouds, the fall of rain, tides in the sea, the shooting of arrows, and so on. Aristotle systematized these obser-

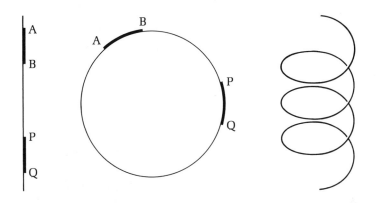

1-5. The straight line and the circle are simple curves in the sense
that any part AB can be superimposed on any other part PQ.
Galileo pointed out that the helical curve on the right also has
this property.

vations by analyzing all the natural examples of motion in terms of
combinations of straight motion and circular motion. What is so
special about straight lines and circles? As shown in Figure 1-5,
these curves are *simple*. Any part of a simple curve can be superimposed on any other part. By using more circles and straight line
bits, it is possible to describe any complicated curve. According to
Aristotle's arguments, this is what happens in nature. Why things
move at all was answered by Aristotle with the statement that each
body has a natural tendency to go to some preferred position, and
the observed motions in nature show bodies moving *in order to get
there.*

Aristotle, however, distinguished this natural motion from another motion, *violent* motion, which is caused by agents like living
beings. Of the examples given above, one, the shooting of an arrow,
is done by a human being, and this motion of the arrow is therefore
not natural but violent. For any such violent motion, there must
always be an efficient cause.

It was not until the seventeenth century that a serious challenge
to Aristotle's ideas was posed. The man to do so was Galileo Galilei

1-6. A painting of Galileo (1564–1642). (Courtesy of Yerkes Observatory.)

(Figure 1-6), mathematician and philosopher to the Grand Duke of Florence. Galileo's genius lay not so much in mathematics and philosophy but in using experimental demonstrations to support his arguments. Galileo's book *Dialogue Concerning the Two Chief World Systems—Ptolemaic and Copernican* is a brilliant demonstration of modern scientific reasoning pitted against the medieval thinking based on Aristotle's philosophy. Not only did Galileo defend the Copernican system, he also attacked the very foundations of Aristotelian natural philosophy.

It will not be possible here to reproduce, even in a brief summary, all of Galileo's arguments and demonstrations against the Aristotelian system. Let us take two examples relating to violent motion as described by Aristotle, shooting an arrow and pushing a cart.

When an arrow is shot from a bow, why does it move? According to Aristotle, there must be an agent acting on it all the time to cause its motion. First, of course, the human being who shot the arrow supplied the cause. But what next? To keep the arrow flying, the Aristotelians had to argue that the air behind the arrow keeps on pushing it, just as wind pushes clouds in the sky. Galileo's reply to

1-7. An arrow shot in the direction of its length goes much farther than an arrow shot perpendicular to its length. Galileo cited this experiment to rule out the Aristotelian principle that things move because air pushes them. In this experiment, air has a larger cross-section of the arrow to push against when it is shot sideways than when it is shot lengthwise.

this reasoning was, first, to show that if an arrow is shot sideways (Figure 1-7), it goes only a short distance. If Aristotle were right would not air have a greater cross-section of the arrow available to push against, and hence to give a greater speed to, than when the arrow is shot lengthwise in the usual manner?

Aristotle had argued that a constant force would generate a constant velocity. To test the correctness of this argument, Galileo constructed a water clock to measure accurate time intervals and then performed the experiment of dropping a heavy body from a great height. If the weight of the body is the force responsible for the fall, according to Aristotle the body should fall through equal

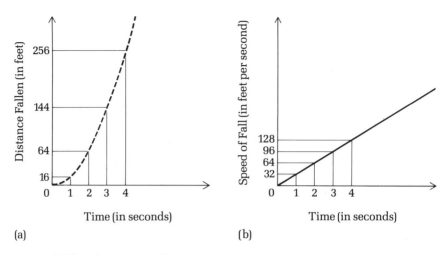

(a)

(b)

1-8. Galileo demonstrated by experiments that, as bodies are dropped from a height, they fall through increasing distances in equal time intervals. As shown by the dashed curve (a), equal intercepts of time correspond to increasing intercepts of distance as the body falls. The speed of fall also progressively increases with time, as shown by the solid straight line (b).

heights in equal time intervals. Galileo demonstrated (Figure 1-8) that the body falls through *increasing* heights in equal time intervals. Its speed does *not* remain constant but *increases* in proportion to time.

When a body is dropped, it is initially at rest and then picks up speed. The same thing happens in other examples in daily life. A cart is at rest. When it is pushed, it starts moving. Thus, what is observed is not a constant velocity but a change of velocity. This change of velocity per unit time is known as *acceleration*. If a car is moving at 50 mph and the driver further presses the gas pedal, the speed of the car increases—it accelerates. Galileo was able to recognize that the real effect when a force is applied to a body is one of acceleration.

Let us go back to the example of the push cart. To cause it to move, we need a force. Stop pushing and the cart stops. So is not Aristotle right when he says that force is needed for motion? Of course, the fallacy in the reasoning begins to show up with a little thought. When we begin to push with a certain force, the cart does not immediately acquire a constant velocity. Its velocity slowly builds—that is, the cart accelerates. When we withdraw the force, the cart should stop at once, if Aristotle were right. However, it continues to move for a while before it stops. And it stops because a force has been acting on it all along, opposing its motion. This is the force of *friction*. So even when the pushing force is withdrawn, the change of speed is caused by a force.

The laws of motion

By arguments and experiments of this kind, Galileo correctly grasped the relationship between force and motion. He realized that force causes a change of motion and that if there is no force acting on a body there will be no change in its velocity. Galileo's appreciation of the relationship between the force and the change of velocity was only a qualitative one. A quantitative statement of this relationship had to wait for a few decades after Galileo. A precise statement of the laws of motion was given by Isaac Newton, who was born the year that Galileo died (1642). In his book *The Mathematical Principles of Natural Philosophy*, published in 1687, Newton (Figure 1-9) gave a detailed discussion of these three laws of motion.

1-9. Portrait of Isaac Newton (1642–1727), from a mezzotint executed in 1740 by James Macardel after a portrait by Enoch Seeman. (National Portrait Gallery.)

Of these, the first law was already known to Galileo. It states that a body will continue to be in a state of rest or of uniform velocity unless some external force acts on it. The type of experiment and reasoning that led Galileo to this law is described in Figure 1-10. Notice the contrast between this law and Aristotle's concept of motion. Aristotle required a constant force to act on a body to generate a constant velocity in it, whereas Galileo's result states that *no* force is acting on the body, as the concept of a constant velocity implies.

Newton's second law states that the acceleration produced in a body is in proportion to the force applied. And here we encounter another notion that was known qualitatively to Galileo, the notion of *inertia*. Qualitatively, it describes the tendency of any piece of matter to resist a change in its state of motion. Quantitatively, we can say that the greater the inertia of the body, the larger the force that is required to produce the same acceleration in it. Greater force is needed to push a car than to push a bicycle because a car has much more inertia than a bicycle. To put the same idea in a different form, for a given force, we can generate a *greater* acceleration in a body of *smaller* inertia. For the same gas consumption, a light car accelerates more than than a heavy limousine.

Newton ascribed a quantitative measure to inertia through the concept of *mass*. Mass is the quantity of matter in a body. The greater the quantity of matter, the greater is the mass and the greater is its inertia. Using Newton's second law, we can compare the masses of two bodies A and B by simply measuring the accelerations produced in them by the same external force. If the same force produces in A an acceleration twice that of B, we conclude that B has twice the mass of A.

Newton's third law of motion states that action and reaction are equal and opposite. If we press against a wall, we feel the wall pressing against us. The force we exert on the wall (action) evokes an equal and opposite force (reaction) from the wall on us.

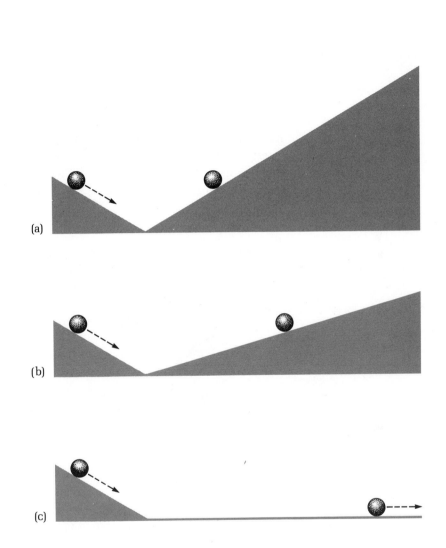

(a)

(b)

(c)

1-10. Galileo arrived at the concept of the first law of motion
through experiments such as the one illustrated here. A ball
is rolled down an inclined plane, and it rolls up another plane
to the same height it started with (a). As Galileo reduced the
inclination of the second plane, the ball traveled farther on it
(b). If the plane is horizontal (c), the ball would continue to
travel forever, in the absence of friction or an intervening
force.

In Figure 1-11, we have a monkey trying to climb a rope over a pulley and carrying a weight equal to that of the monkey. The weight is at the same distance from the pulley as the monkey is. Where will the weight be by the time the monkey reaches the top? Use Newton's third law and deduce that the weight will also reach the top at the same time that the monkey does!

1-11. A monkey attempts to climb a rope. Whatever pull he exerts on the rope to draw himself up is communicated to the stone by the law of equality of action and reaction. The stone therefore moves up in the same way the monkey does.

Some concepts in dynamics

Before we end this chapter, let us define some concepts of *dynamics*, the subject dealing with the motion of bodies under different forces according to the three Newtonian laws of motion.

We have already mentioned *velocity* and *acceleration*. The concept of velocity actually includes two bits of information, how fast the body is moving and in what direction it is moving. The first bit of information signifies *speed*. Thus, the information that a car is moving at 60 mph tells us about the speed of the car. To know its velocity, we must know also the direction in which it is proceeding.

Acceleration, as mentioned before, denotes the *rate of change* of velocity. A change in the velocity could occur in two ways, through a change in the speed or a change in the direction. In Figure 1-12, we see a car going around a circular race track with a constant speed of 150 mph. Although its speed is constant, its direction is changing all the time as it goes in the circle. Therefore the car is accelerating.

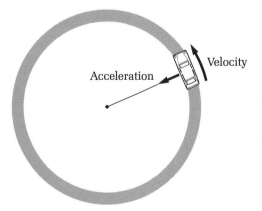

1-12. A car moving in a circular track has a continuously changing direction. It may have a constant speed, but its velocity changes because of the change of direction. The car has a resulting acceleration toward the center of the track.

Like velocity, acceleration also has magnitude and direction. What is the acceleration of the car in this case? The answer is usually obtained with the aid of calculus, but it can be stated in a simple form. The acceleration is directed toward the center of the race track and has a magnitude equal to the square of the speed of the car divided by the radius of the track. If the track has a one-mile radius, the magnitude of the acceleration is 150 mph × 150 mph = 22,500 mphph.

Another useful concept is that of *angular velocity*. In the example of the race car, when it has made one complete round, we say it has completed a total angle of 360° about the center. How long does the car take to make one round? We already know that the speed of the car is 150 mph and the circumference of the track is 2π miles.* Therefore, the time taken to make one round of the track is $2\pi/150$ hours, or about $2\frac{1}{2}$ minutes. Since during this time the car traverses an angle of 360°, its angular velocity has the magnitude of about $360°/2\frac{1}{2}$, or 144° per minute.

Let us now look at the scenario shown in Figure 1-13. This sequence of events could easily have come from a Laurel and Hardy comic movie. A butterfly hits Laurel, causing him a mild annoyance but no more trouble. However, when Hardy, in hot pursuit of the butterfly, hits Laurel, the effect cannot help being spectacular. What property of motion is crucial in producing effects like these?

For the purpose of this illustration, we may assume that Hardy and the butterfly had the same speed, but their effects on Laurel are not the same. Clearly, the large mass of Hardy has made all the difference. But this is not all! If Hardy were walking slowly, he would not have bumped into Laurel so hard. The total effect is due therefore to mass as well as the velocity. The quantity that combines the two is called the *momentum*. Momentum is simply the product of mass and velocity. It has the same direction as that of the velocity.

---

*The circumference of a circle is 2π times its radius. The constant $\pi = 3.141596\ldots$, and it is often approximated by the fraction 22/7.

1-13. This comic strip illustrates how the momentum of Hardy is communicated to Laurel during their collision.

If we go back to Newton's second law, we now see that it could also be stated in the following form: The rate of change of momentum is equal to the force applied. In an impact with Laurel, Hardy's momentum has clearly changed, the change being caused by the force of impact. And, as Newton's third law implies, Laurel would feel an equal and opposite force of reaction, which is why he is thrown off his chair.

A corollary of Newton's second law is that if there is no net force on a body (or a collection of bodies), the total momentum is unchanged. In the Laurel–Hardy collision, there is no net force—the equal and opposite forces of impact cancel each other and so their total momentum is unchanged. Before impact, Laurel was at rest while Hardy was moving. Afterwards, most of the momentum is carried by poor Laurel. This unchangeability of total momentum is called the law of *conservation of momentum.*

A related concept is that of *angular momentum*, and to understand it let us go back to the example of the car on the race track. Suppose the mass of the car is 2000 lbs. Then its momentum is

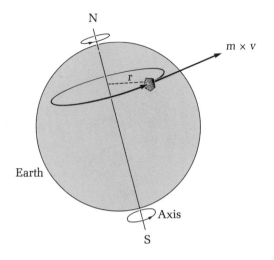

1-14. A typical bit of matter inside the rotating Earth goes around the axis of rotation (the north–south axis) at a distance r. If v is the speed of this bit and m is its mass, it has a momentum m × v in the tangential direction. The angular momentum of this bit about the axis of rotation is m × v × r. The angular momentum of the Earth is obtained by adding the angular momentum of all the bits that make up the Earth.

simply 2000 (lbs.) × 150 (mph) = 300,000 lbs. mph. Its angular momentum about the center of the track is given by multiplying this momentum by the radius of the track.

In general, a rotating body possesses angular momentum. In Figure 1-14, we see the Earth rotating about its axis. How do we calculate its angular momentum? To do so, we divide the Earth into tiny bits (much as one would divide a picture into jig-saw bits). In Figure 1-14, we see a typical bit moving in a circular track whose center lies on the axis of rotation. We multiply the momentum of this bit by the radius of the circle along which it moves, just as we did for the car on the race track. We add the contributions from all such bits to get the total angular momentum of the Earth. Like the law of conservation of momentum, we also have a law of *conservation of angular momentum* of a system, provided there are no net forces on the system with a tendency to affect its angular motion.

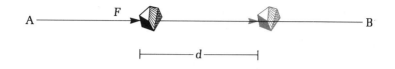

1-15. The work done by the force *F* in moving the object over a distance *d* from A to B is *F* × *d*. Only the displacement along the direction of the force counts as work.

The constancy of angular momentum plays an important role in the dynamical evolution of many astrophysical systems.

Finally, we consider the important concepts of work and energy. The arguments given by Galileo and Newton have already clarified the relationship between force and motion. Contrary to what Aristotle said, we now see that forces act to change the state of motion rather than to maintain it at a constant velocity. Is there any method of bookkeeping that can tell us what has been achieved by the applied forces at any stage of the motion?

Physicists have found a way of mathematically defining the *work* done by such forces. In Figure 1-15, we see that, by the application of a constant force *F*, a body has been displaced from position A to position B. Let *d* denote the net displacement of the body *along the direction of the force F*. Then the product *F* × *d* denotes the *work done by the force*.

Is there any manifestation of this work? We know that the body has accelerated as a result of the application of a force. Suppose it were at rest at *A* and has attained the velocity *v* at *B*. A simple calculation using Newton's second law of motion will show that the work done by the force *F*, defined earlier as *F* × *d*, is exactly equal to $\frac{1}{2}m \times v \times v$, or $\frac{1}{2}mv^2$, where *m* is the mass of the body.

This quantity, $\frac{1}{2}mv^2$, is the *kinetic energy* of the body—the energy it acquires by virtue of its motion. So we see that the work done by the forces has not gone in vain but has resulted in giving an equal amount of kinetic energy to the body.

If, as often happens in the case of frictional forces, the external force opposes the motion of the body, the body slows down. Here we have a case of *reduction* in the kinetic energy of the body.

The expenditure of energy has gone toward work done *against* the external force. Thus we can state a general rule that applies to moving bodies:

change in the kinetic energy = work done by an external force

If the change is negative (corresponding to a reduction of the kinetic energy), the work is also negative, that is, against the external force. In Chapter 3 we will relate this rule to the law of *conservation of energy.*

Why things move

This brings us to the end of our discussion of dynamics, the science of motion, perhaps the oldest of all sciences. Why do things move? An answer to this question was first attempted by Aristotle and was later given in the correct form by Galileo and Newton. Galileo was the first to realize that the effect of forces lies in *changing* the state of motion, while Newton gave quantitative expression to the relationship between force and acceleration.

In nature we see so many different types of forces. Yet with painstaking advances in theory and experiments, scientists have reduced the basic categories of these forces to only four. In this book we will discuss examples of all four types of basic forces, although our main emphasis will be on gravity, the strangest force of them all.

The Apollo 11 Saturn V rocket carried astronauts Neil Armstrong, Michael Collins, and Edwin Aldrin on the first manned journey to the Moon. (NASA.)

2

From the Falling Apple to Apollo 11

2

From the Falling Apple
to Apollo 11

Why did the apple fall?

Apples have played a prominent role in many legends, myths, and fairytales. It was the forbidden apple that became the source of temptation to Eve and ultimately brought God's displeasure upon Adam. It was the apple of discord that led to the launching of a thousand ships and the long Trojan War. It was a poisoned apple that nearly killed Snow White, and so on.

For physicists, however, the most important apple legend concerns the apple that fell in an orchard in Woolsthorpe in Lincolnshire, England, in the year 1666. This particular apple was seen by Isaac Newton, who "fell into a profound meditation upon the cause which draws all bodies in a line which, if prolonged, would pass very nearly through the centre of the earth."

The quote is from Voltaire's *Philosophie de Newton* published in 1738, which contains the oldest known account of the apple story. This story does not appear in Newton's early biographies, nor is it mentioned in his own account of how he thought of universal gravitation. Most probably it is a legend.

It is interesting to consider how rare it is to see an apple *actually fall* from a tree. An apple may spend a few weeks of its life on the tree, and after its fall it may lie on the ground for a few days. But how long does it take to fall from the tree to the ground? For a drop of, say, 9 feet, the answer is three-quarters of a second. So to see an apple fall, we have to be on hand during this crucial short interval of its life! The chance of witnessing this event of course increases if we sit in an apple orchard in the right season, but still, the event as such cannot be considered very frequent.

Much less frequent is the appearance of a genius like Newton, who could meditate on such an event and come up with the law of gravitation. Legend has it that Newton's meditations on the question "Why did the apple fall?" led him eventually to the inverse-square law of gravitation. Newton's answer to this question— "because the Earth attracted it"—is more profound than it appears to be at first sight, for it not only resolved the mystery of the falling apple but helped resolve a number of long-standing questions about our solar system.

What is the law of gravitation?

Stated in simple words, the law of gravitation tells us that the force of attraction between any two material bodies increases in direct proportion to their masses and decreases in inverse proportion to the square of their distance apart. In the compact language of mathematics, this law is stated

$$F = G \times \frac{m \times M}{d^2}$$

In this formula, F is the force of gravity between two bodies of respective masses m and M situated at a distance d apart, and G is a universal constant. We have already encountered the term *mass*; it is defined as the quantity of matter in a body, and it is also a measure of a body's inertia. We now find another meaning ascribed to it; mass is a measure of how strongly a body can exert a gravitational force on other bodies and a measure of how susceptible a body is to the gravitational influence of other bodies. In Newton's formula, if we increase m by a factor of 10, the force F is correspondingly increased by a factor of 10. If we decrease m by a factor of 10, the force F is correspondingly decreased by a factor of 10. Because of this property, gravity appears to play a negligible role in the behavior of atoms and molecules, which have very small masses, whereas it becomes an important force in astronomy, a subject dealing with heavenly bodies of very large masses.

Because of gravity's diminishing influence with distance, this law is often referred to as the *inverse-square law*. This inverse-square relationship is common in nature. For example, it also ap-

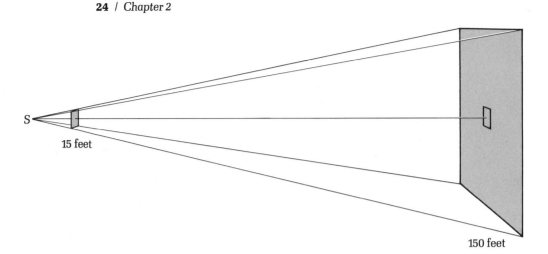

15 feet

150 feet

2-1. A source of light S viewed from 10 times farther away appears
100 times fainter. The amount of light falling on the small
square 15 feet away is equal to the amount of light falling on
the large square 150 feet away. The area of the large square is
100 times the area of the small square. A small square of the
same size at 10 times the distance receives only one-hundredth
of the light that the nearby square receives.

plies to the amount of light we receive from a luminous body. If
we look at a 100-watt light from a distance of 15 feet, we find it to
be very bright. The same light viewed from 150 feet appears faint.
Consider a fixed area, as in Figure 2-1, held perpendicular to the
path of the light rays. When we increase our distance from the light
source by a factor of 10 (from 15 feet to 150 feet), the amount of light
we collect on this area per second is reduced by a factor of 100
(10^2). The same relationship occurs in the case of F, the force of
gravity. If we increase the distance d by a factor of 10, the force F
is diminished by a factor of 10^2, or 100.

At this point, it is worth asking "Why should gravity be important in astronomy and negligible in atomic physics, when the distances are large in the former and small in the latter?" The answer
is that, although according to the inverse-square law the force of
gravity should be strong at atomic distances, it is overshadowed by
other forces that are considerably more powerful. For example, the
force of *electrical* attraction between the electron and the proton in

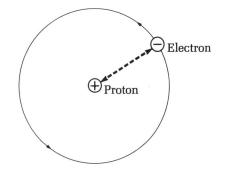

2-2. The electrical attraction between an electron
and a proton in the hydrogen atom is about
10,000,000,000,000,000,000,000,000,000,000,000,000,000
times their gravitational attraction toward each other.

a hydrogen atom (Figure 2-2) is estimated to be something like ten
thousand billion billion billion billion (10^{40}) times the force of their
gravitational attraction! Inside the nucleus of the atom, the nuclear
forces are even stronger than the electrical forces. The atomic
physicist therefore rightly ignores gravity in his calculations.

In astronomy, however, none of the other competing forces of
nature remain to challenge gravity. The nuclear force is of a very
short range; its effect dies out beyond distances of a thousand-
billionth of a centimeter! Since the heavenly bodies are electrically
neutral, their electrical force of attraction is zero. This is why
astronomers find gravity to be the most important force in their
calculations.

Newton argued, on the basis of the law of gravitation, that the
apple fell because it was *attracted* by the Earth's gravitational
pull. But how did he arrive at the form of the inverse-square law of
attraction? Surely, if the purpose of the law was to explain only the
falling apple, then it could be served by any law of attraction!
In fact, what led Newton to the inverse-square law was not the
need to explain the falling apple but the need to explain a much
bigger phenomenon—the motion of planets and satellites of our
solar system.

The motion of planets

In Chapter 1, we saw how Aristotle's ideas had dominated scientific thinking right up to medieval times. Aristotle's ideas led to the so-called *geocentric* theory of the Universe, which assumed that the heavens revolved around the fixed Earth, thus explaining why the Sun and the stars systematically rise in the east and set in the west. We have already mentioned how Aristotle's ideas came to be regarded as absolute truths not only in Greece but throughout Europe. The following example from India illustrates how the influence of the geocentric theory had spread well beyond Europe.

The notion of a fixed Earth and the moving cosmos was challenged in India by Aryabhata, a distinguished astronomer of the fifth century AD. In his Sanskrit text on astronomy, the *Aryabhateeya*, there is an explicit mention of the Earth's rotation about its axis (chapter 4, verse 9): "Just as a man rowing a boat sees the trees on the bank of the river go in the opposite direction, so do the fixed stars appear to us to move from east to west." However, so well rooted was the "fixed Earth–moving cosmos" in Indian astronomy that Aryabhata's pupils and successors either denied that he ever held such contrary views or tried to interpret the above verse differently to make it appear less offensive to contemporary scholars.

The geocentric theory, however, did not stop at the motions of stars. The star motions were quite regular and fell within Aristotle's criterion that natural motions are circular or in straight lines. There was another class of objects, the planets, whose motions were considerably irregular. (The Greek word *planet* means wanderer.) Some planets, like Venus, showed *retrograde* motion (see Figure 2-3), whereas others appeared to decrease or increase in speed at times. To accommodate such motions into the Aristotelian system, the Greek astronomers, especially Hipparchus and Ptolemy, made elaborate geometrical constructions involving circular paths known as *epicycles*. In this effort, they were successful to the extent that their *epicyclic theory* could predict in what part of the sky a planet would be found at a given date. The

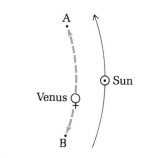

2-3. As seen from the Earth (⊕), the Sun (☉) goes around in the direction of the long arrow. Venus (♀) also appears to go around, but it sometimes goes ahead of the Sun (to A) and sometimes falls back (to B). This latter movement is known as retrograde motion.

demands on observational accuracy in those days were not so rigorous as they are today, and the successes of this theory naturally gave it the status of a dogma.

The geocentric theory was challenged by Nicolaus Copernicus (1473–1543), who proposed a rival framework in which to describe the motions in the solar system. This framework, known as the *heliocentric theory*, assumes the Sun to be fixed in space, and the planets, including the Earth, to be orbiting around it. Like Ptolemy, Copernicus also gave elaborate constructions involving circles (an influence of Aristotle?) to describe planetary motions (see Figure 2-4).

The Copernican constructions are simpler but no more accurate than those given by the old geocentric theory. However, their main merit lies in the fact that these constructions, for the first time, pinpoint the central place of the Sun in the planetary system. For someone looking for a dynamical theory—for an explanation of why planets move—the importance given to the Earth in the geocentric theory would be misleading. The clue to the motion of planets lies, as we shall see later, not in the Earth but in the Sun.

The Copernican hypothesis received considerable opposition during Copernicus' lifetime. Copernicus did not see the published version of his book *De Revolutionibus Orbium Caelestium* until he

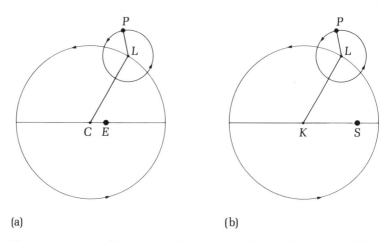

(a)   (b)

2-4. The construction of Ptolemy is shown in (a). The Earth is at E and the planet at P. P moves on a circle with center at L. The center L in turn moves on another circle around E but not centered at E. Ptolemy gave elaborate prescriptions for the dimensions of these circles as well as the rates at which the points P and L move on their respective circles. The Copernican construction shown in (b) also involves circles, but now the Sun is identified as the fixed point S. The planet P moves on a circle whose center L moves on another circle *not* centered on S.

was on his deathbed. However, its impact on succeeding generations, though gradual, was tremendous.

We have already seen in Chapter 1 how the Copernican theory received strong support from Galileo. It was Johannes Kepler (1571–1630), however, whose painstaking observational work marked the next improvement over the Copernican theory. Copernicus had attempted to use circles to describe orbits of planets, but Kepler discovered that these orbits are best described by *ellipses*. Kepler arrived at the following three laws of planetary motion (see Figure 2-5 for illustration of these laws):

1. *The orbit of a planet is an ellipse with the Sun as one of its two foci.*

2. *The radial line from the Sun to the planet sweeps out equal areas in equal intervals of time.*

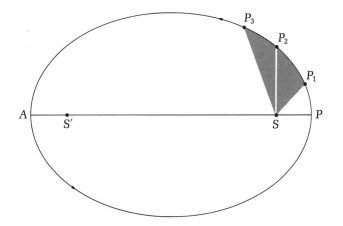

2-5. The orbit of a planet is best described by an elliptical curve. The Sun S is at one of its two foci, the other focus being at S'. The planet moves along the ellipse in such a way that the line joining it to the Sun sweeps out equal areas in equal times. Thus, if the planet goes from P_1 to P_2 and from P_2 to P_3 in equal intervals, then the areas SP_1P_2 and SP_2P_3 must be equal. The line AP is called the *major axis* of the ellipse, with A (the farthest point from the Sun) known as the *aphelion* and P the point nearest the Sun) known as the *perihelion*.

3. *The square of the time taken by the planet to complete one orbit varies in proportion to the cube of the major axis of the orbit.*

Kepler's laws provided the empirical background to Newton's dynamical theory. Kepler's laws described *how* planets move; Newton's laws of motion and gravitation supplied the reason *why* the planets move according to Kepler's laws.

Newtonian gravity and motion in the solar system

To draw a circle with a radius r centered at S, we attach one end of a string to S and the other end to a drawing pencil P. The length of the string between the two ends is r. Keeping the string taut, we

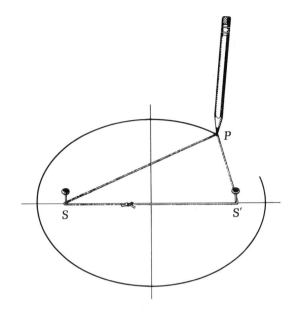

2-6. How to draw an ellipse. The ratio of the distance SS′ to the length of the string PS + PS′ is called the *eccentricity* of the ellipse. When the eccentricity is zero, S and S′ coincide, and the ellipse becomes a circle.

move the pencil around, and it draws a circle. How do we draw an ellipse with foci at S,S′ and semimajor axis a? The construction is a little more elaborate (see Figure 2-6). Take a piece of string of length $2a$ and attach its ends to S and S′. Move a pencil with its end P sliding across the string such that the bits PS and PS′ are always taut. In the case of a circle, the pencil end always maintains the distance PS = a; for the ellipse, we have PS + PS′ = $2a$. For the construction of the ellipse, the distance SS′ clearly cannot exceed $2a$. If S and S′ coincide, the ellipse becomes a circle.

Newton used his system of dynamics to describe the motion of planets pulled by the Sun's gravity. His equations of motion (see Chapter 1) relate the acceleration of the planet to the impressed force, in this case, the force of gravity. Knowing the planet's acceleration, can we calculate its actual path in space? To solve this problem, Newton developed a new branch of mathematics which he called *fluxions* but is now known as the *calculus*. The methods of calculus enabled him to prove that the planets move along elliptical paths satisfying Kepler's three laws. However, the scientific

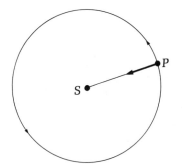

2-7. The derivation of the inverse-square law of gravitational
attraction follows from Kepler's laws of planetary motion.
The derivation of the former from the latter is particularly
easy in the simplified situation where the planet P moves in a
circular orbit with the Sun S at the center. The force and the
acceleration to which P is subjected by S are both along the
radial direction shown by the arrow.

community always tends to be conservative and views new methods with suspicion. To make his theory more readily acceptable,
Newton therefore recast his simple proofs based on the calculus
into the more conventional but more cumbersome geometrical
forms. Newton's book *Philosophiae Naturalis Principia Mathemat-
ica* published in 1687 contains his momentous work on motion and
gravitation.

Without going into intricate mathematical details, it is possible
to see why an inverse-square law of gravitation is implied by
Kepler's observations. We will consider the simplified problem of
motion in a circle, which we saw above to be a special case of the
ellipse.

In Figure 2-7, we have a planet P of mass m moving in a circle
about the center S where the Sun is located. First, we should note
that, if the radial line SP is to sweep out equal areas in equal
intervals of time (Kepler's second law), P must move with a constant speed along this circle. Suppose the radius of the circle is r;
then its circumference is 2πr. If the time taken by the planet to go
once around this circle is T, then its constant speed v must be

$$v = \frac{2\pi r}{T}$$

In what direction must a force act on P for it to move in a circle? To argue that the force must be *in the direction of motion* is to make the same mistake that Aristotle and his followers made. Force is related *not* to the velocity but to the acceleration. And the acceleration of P is toward the center S and is equal to v^2/r (see Chapter 1). Thus the force F on the planet must be toward the center S and is given by Newton's second law, Force = mass × acceleration, or

$$F = m \times \frac{v^2}{r}$$

Since we know v as $2\pi r/T$, we get

$$F = m \times \frac{(2\pi r/T)^2}{r} = \frac{4\pi^2 mr}{T^2} .$$

Now we use Kepler's third law, which tells us that T^2 increases in proportion to r^3, or

$$T^2 = k\, r^3$$

where k is a fixed number. Substituting for T^2 in the expression for the force F, we get

$$F = \frac{4\pi^2 mr}{kr^3} = \frac{4\pi^2 m}{k} \times \frac{1}{r^2}$$

This tells us that the force on the planet P decreases in inverse proportion to the square of its distance from S; that is, it varies in accordance with the *inverse-square law!*

The law of gravitation not only describes the motion of planets around the Sun but also the motion of the Moon around the Earth and the motion of other satellites around their respective planets. That the same law describes the falling apple and the Moon's motion may seem surprising at first. That the moon is *continually falling* on the Earth (as the apple did) is seen with the help of Figure 2-8. There the Moon M is shown as moving in a circle around the Earth E. Suppose, by magic, we switch off the force of attraction of the Earth. As shown in the figure, the Moon would then start moving along the dotted straight line with a uniform speed— because there is no force acting on it (Newton's first law)! Compare this trajectory with the Moon's actual circular trajectory around the Earth. Left to itself, the Moon's natural tendency is to break

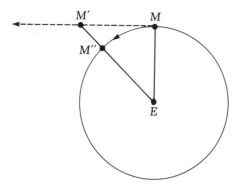

2-8. In the absence of the gravitational pull of the Earth, the Moon
would have moved from M to M' along the dashed straight
line. Because of gravity, the Moon actually moves along the
circular arc MM''. Relative to the Earth E, the circular arc is
closer than the straight line. Thus the Moon, as it moves
in the transverse direction, also falls *toward* the Earth
(in this case, from M' to M'').

away along the dotted trajectory; but the Earth is constantly
pulling the Moon toward it. Hence we can look upon the Moon's
motion as if it is continually falling toward the Earth. Because it
has transverse velocity, it never actually reaches the Earth but
keeps moving sideways.

Who first thought of gravitation?

Newton was not the first to have thought of gravitation. As early
as the fifteenth century, some astronomers had the idea that a force
of attraction might exist between heavenly bodies and the Earth.
It was argued that the Earth is being pulled in all directions by a
"magnetic" force, but since the force is the same in all directions,
the Earth remains at rest.

Gilbert in 1600, Ismaelis Bouillard in his book *Astronomia Philo-
laica* published in 1645, and Alfonso Borelli in 1666 appear to
have come close to the basic features of the Newtonian law of
gravitation, as did Kepler, who once actually considered the
inverse-square law before rejecting it.

2-9. Robert Hooke (1635–1703). (From *Lives in Science*. Copyright © 1957 by Scientific American, Inc. All rights reserved.)

The apple legend credits Newton with the idea of gravitation in 1666, although his first publication on it, a treatise called *Propositions de Motu,* was communicated to the Royal Society in February 1685, while the *Principia* itself was published in 1687. In the meantime, in 1674, Robert Hooke (Figure 2-9) published his work describing the motion of the Earth around the Sun in terms of a law of attraction that decreased with distance. Hooke is said to have communicated his ideas to Newton, who had also arrived at similar conclusions independently.

Why did Newton wait for so long—nearly two decades—before publishing his results? In the present scientific era of "publish or perish," where rushing to the media for announcement of half-baked results is not uncommon, it becomes all the more difficult to understand Newton's reticence.

It is argued that Newton was perfectionist and wanted to wait until he had sorted out some problems connected with his theory. One of these problems was the need of a mathematical proof that a spherical body attracts others as if its mass were concentrated at its center (see Figure 2-10). The other problem was an observa-

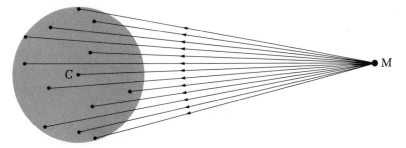

2-10. If we divide a sphere into tiny bits, each bit will gravitationally attract a particle *M* toward itself. A few such bits are shown in the figure. The result of *all* these forces on *M* will be a force acting in the direction *MC*, where *C* is the center of the sphere. The magnitude of the force will be what it would be if the entire mass of the sphere were concentrated at *C*.

tional one. It seems that Newton wanted to wait until reliable estimates of the dimensions of the Earth–Sun–Moon system became available so that he could test the correctness of his theory. These became available in the late 1670s. It was only then that Newton felt confidence in his law of gravitation.

Controversy still exists about why Newton waited and about the extent of credit that Hooke should be given. About the finished product, however, there is no doubt. The credit for the working of planetary orbits based on the laws of motion and gravitation goes to Newton. None of his contemporaries had the mathematical expertise or the breadth of knowledge to carry through such a calculation.

Newton's dislike for controversy and his reticence are reflected in his communication to Halley when submitting Book II of the *Principia* for publication. By then, Book I of the *Principia* had been published, and Hooke had claimed priority in the authorship of the law of gravitation. Halley was acting as the peacemaker in the controversy. Referring to Book III (to follow Book II), Newton wrote: "The third I now design to suppress. Philosophy is such an impertinently litigious lady that a man had as good be engaged in law-suits to have do with her." However, to the advantage of posterity, Halley succeeded in persuading Newton to change his mind.

Successes of the law of gravitation

Leaving aside the controversy about who should be credited with the genesis of the law of gravitation, let us now review some of its achievements. The law of gravitation implied instantaneous action at a distance. The force of gravity between the Sun and the Earth is communicated instantly across a distance of some 90 million miles. How is this done? Why did the attraction diminish according to the inverse-square law? Questions like these did trouble Newton's contemporaries and successors. When asked about such questions, Newton is believed to have said "Non fingo hypotheses" (I do not invent hypotheses). Newton attached more importance to the requirement that the law should adequately describe observations than to the deeper questions of nature's mysterious processes leading to that law.

Indeed, it was the successes achieved by Newton's law that established it so firmly in post-Newtonian physics. The bothersome deeper questions of how and why were relegated to the background by the successes of the inverse-square law. Let us look at some of its triumphs.

The first example is that of Halley's Comet (see Figure 2-11). Like a planet, the comet also moves in orbit by the Sun's gravitational pull. However, unlike a planet, the comet moves in a highly eccentric orbit. If we go back to our construction of an ellipse, we recall that a highly eccentric ellipse will result if the separation between S and S' is very nearly equal to (but still less than) the length $2a$. An example of a cometary orbit as distinct from that of a planet is shown in Figure 2-12.

As a result of moving in such an orbit, the comet is seen in the vicinity of the Sun after a long duration. But, unless the orbit of the comet (which extends to remote parts of the solar system) is disturbed by an intervening planet like Jupiter, its visits in the neighborhood of the Sun are periodic.

Edmund Halley, a contemporary and friend of Newton, noticed this periodicity for a comet that was seen in 1682. Halley argued that this was the same comet that had come earlier in 1456, 1531, and 1607—at a regular interval of a little over 75 years. Halley predicted that it would be seen again in 1758. This prophecy came true, although Halley did not live to see the comet's passage in that year. Halley's Comet is next due to return in our lifetime during 1985–1986.

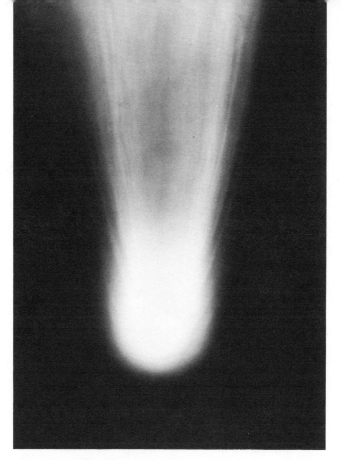

2-11. Photograph of Halley's Comet.
(Courtesy of Mount Wilson and Las Campanas Observatories,
Carnegie Institution of Washington.)

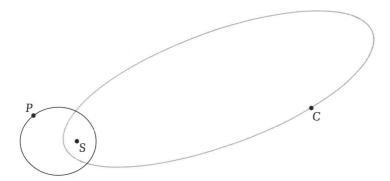

2-12. Typical orbits of a planet P and a comet C. Both orbits are
ellipses with the Sun S at the focus. The planetary orbit is
nearly circular, while the orbit of a comet is a highly eccentric
ellipse.

Perhaps no one did more to establish confidence in the law of gravitation than Pierre Simon Laplace (1749–1827), the French mathematician. Laplace's five-volume work *Mecanique Celeste* published from 1799 to 1825 has been compared with Ptolemy's *Almagest* for its impact on contemporary astronomy. In this work, Laplace applied the latest mathematical techniques to work out the motions of planets and their satellites under each others' gravitational influence. The problem is extremely intricate when one takes into account all the cross-influences of the eighteen bodies (then known) of the solar system. Faced with such a problem in modern days, the inclination of the physicist would be to "put it all on a computer." The success achieved by Laplace in solving his mammoth problem, and the resulting agreement between his calculations and the observations of planets and satellites, convinced the skeptics about the validity of Newton's law of gravitation. When Napoleon asked Laplace why his book made no mention of God, he is said to have replied, "I had no need of that hypothesis."

The next triumph of Newtonian theory came in 1845, when it was used in the discovery of a new planet. Two astronomers, Adams in England and Le Verrier in France, came to this discovery working independently. Their work was based on the irregularity that had been noticed in the orbit of Uranus, then the farthest-known planet of the solar system. Uranus was apparently not following the elliptical orbit predicted by Newtonian gravity. Both Adams and Le Verrier concluded that the irregularity in the motion of Uranus was caused by a new planet in its vicinity; the gravitational pull of this new planet would be responsible for the perturbation of Uranus' orbit. The two astronomers were able to calculate where the new planet should be located. Whereas the suggestions from Adams to observe this planet were ignored by Challis and Airy, the leading observational astronomers in England, Le Verrier's work was taken seriously by Galle of the Berlin Observatory, and he did succeed in locating the new planet Neptune. The Neptune episode illustrates the fact that, if a scientific theory follows the right lines, apparent disagreement with its predictions can lead to new scientific discoveries.

While these three examples relate to the natural constitutents of the solar system, our fourth and last example deals with artificial satellites and spaceships. The motion of these objects—whether it

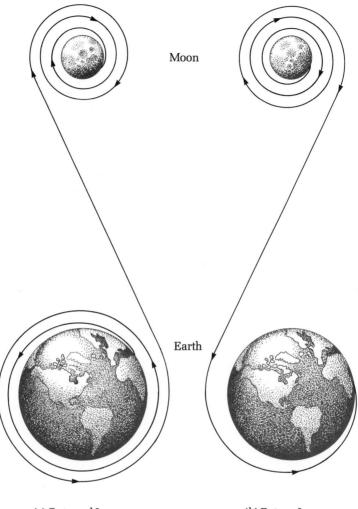

(a) Outward Journey   (b) Return Journey

2-13. Schematic diagram of the orbit of a spaceship from the Earth
to the Moon and back. The dynamics of this complicated
series of motions is governed by Newton's laws.

be the first Sputnik to go around the Earth, or the Apollo 11
mission to the Moon, or the Viking, Pioneer, and other space
missions to other planets—is governed by the same law of gravitation that Newton gave three centuries ago (see Figure 2-13).

For example, the Apollo 11 journey from the Earth to the Moon (and back!) had to take into account the following motions. First, there is the Earth's motion around the Sun and the Moon's motion around the Earth. In fact, we have a "three-body problem" in which each body moves under the gravitational pulls of the other two. Next, the motion of the spaceship from the Earth to the Moon is governed by the gravitational pull of the Earth and the Moon on the spaceship. The calculation of the correct trajectory is complicated and can be done with ease only on an electronic computer.

The accuracy of present-day space missions is considered to be a triumph of modern technology. It is also a vindication of the law of gravitation, allegedly inspired by a falling apple. It is therefore with some confidence that we next consider even more remarkable manifestations of gravity in astronomy.

Astronaut Edward H. White II was the first American astronaut to leave his spacecraft while in orbit. White experienced the weightlessness of space during the Gemini–Titan 4 Mission in 1965. (NASA.)

3

How Strong Is Gravity?

3

How Strong Is Gravity?

The mass of the Earth

Atomic physicists consider gravity to be the weakest of the four known basic forces of nature. Yet, astronomers find gravity to be the most dominant force in the celestial environment. How do we assess the *strength* of gravity in any given situation? We will try to answer this question with a few examples in this chapter.

All of us on the Earth are conscious of gravity. The feeling of weight that we have results from the gravitational force the Earth exerts on us. Newton's inverse-square law of gravitation tells us how strong this force is on any given body on the Earth's surface. Let m be the mass of the body and M the mass of the Earth. Newton's law then tells us that the force of attraction between the body and the Earth is given by

$$F = G \times \frac{m \times M}{R^2}$$

where G is the constant of gravitation (see Chapter 2) and R is the distance between the body and the Earth.

What is R? As shown in Figure 3-1, the Earth is nearly a perfect sphere with a radius of about 4000 miles. From where on Earth should we measure the distance to the body? The distance is zero from the point on the surface where the body lies, while it is 8000 miles from the diametrically opposite point. The distance of any other point in the Earth will lie somewhere between these extreme values.

Here we recall a result from Chapter 2, a result that is said to have taken Newton several years to prove: *A spherical body attracts as if its entire mass were concentrated at its center.* For a spherical

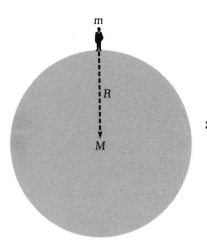

3-1. A body of mass *m* close to the surface of the Earth is attracted as if the entire attracting mass *M* of the Earth (assumed to be spherical here) is concentrated at the center of the Earth.

Earth, the correct value for *R* is therefore equal to the Earth's radius—about 4000 miles.

We now come to an interesting consequence of the above result. If we know the value of *G*, the constant of gravitation, we can measure *the mass of the Earth!* Henry Cavendish (1731–1810) first gave an experimentally measured value of *G*. Figure 3-2 shows a portrait of Cavendish with his measuring apparatus. Of course,

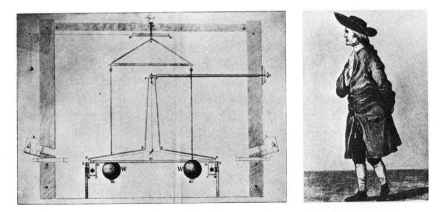

3-2. Henry Cavendish (1731–1810), whose portrait appears at right, devised an experiment to measure *G*, the constant of gravitation. In Cavendish's apparatus (shown at left), the gravitational force exerted by two massive balls on two hanging pellets is measured by balancing it with a torsional force, that is, the force that arises from the tendency of a twisted string to unwind itself.

3-3. Galileo dropped objects of various shapes and sizes from the top of the Leaning Tower of Pisa to demonstrate that all bodies fall with equal speed. The tower is 179 feet high and has a maximum inclination from the vertical of about 13 feet, or just over 4°.

modern techniques give us the value of G much more accurately than the early measurements of Cavendish. The value is 6.66 × 10^{-8} in units of cubic centimeters per second squared per gram ($cm^3/sec^2/g$).

Let us see how we can measure M, the mass of the Earth, with this information. First, recall the experiment performed by Galileo at the Leaning Tower of Pisa (Figure 3-3). Galileo demonstrated that all bodies dropped from a height fall with equal velocity. This experiment is easily interpreted in terms of Newton's second law of motion, Force = Mass × Acceleration. We have already seen what the force of gravity is on a body on or close to the surface of the Earth. Dividing that force by the mass of the body gives us the acceleration. Denoting acceleration by the symbol g, we get the simple formula

$$g = G \times \frac{M}{R^2}$$

Notice that g, the acceleration of the falling body, *does not* depend on its mass, which explains Galileo's conclusion that all bodies fall with equal velocity.* The value of g is approximately 32 feet per second per second.

We can write this relation in a slightly different form:

$$M = \frac{g \times R^2}{G}$$

When written this way, we have *all known quantities* on the right side of this relation: g, the acceleration of the falling body; R, the radius of the Earth; and G, the constant of gravitation. A simple calculation then gives us the *mass of the Earth* as 6×10^{24} kilograms, approximately. The enormous size of this mass by our everyday standards should clarify why the force of gravity is felt by all of us. The Earth is the single most massive object in our environment.

This same method used for measuring the mass of the Earth can also be used for measuring the mass of the Sun. In Chapter 2, we saw that the planets moving around the Sun are also continually falling toward it. It is therefore possible to think of the acceleration of the Earth as it "falls" around the Sun. Thus, the formula that enabled us to measure the mass of the Earth can now be used to estimate the mass of the Sun. All we have to do is substitute for g the acceleration of the falling Earth and for R the radius of the Earth's orbit. The mass of the Sun turns out to be about 2×10^{30} kilograms. Thus the sun is about 330,000 times as massive as the Earth!

Mass and weight

A common mistake in everyday language sometimes causes confusion between the concepts of *mass* and *weight*. Mass is the quantity of matter contained in a body. As we saw in Chapter 1, mass is the measure of a body's inertia, the property by virtue of

---

*In arriving at this conclusion, *air resistance* as a force is neglected. A sheet of paper dropped from a second-story window does *not* fall as rapidly as a pen dropped from the same place because the air resistance is considerably higher on the paper than on the pen. Galileo was aware of this effect.

which the body resists any change in its existing state of rest or motion. In Chapter 2, we found another property of mass—it measures the strength with which the body attracts and is attracted by other bodies. These properties of mass will continue to apply to the body without any quantitative change *no matter where it is in the universe;* mass is an *intrinsic* property of the body.*

The *weight* of a body, on the other hand, measures the *force* with which it is gravitationally attracted by the Earth or by any other nearby body. Weight can therefore vary, depending on the location of the body. Even on the Earth, the weight of the same body can vary from place to place. Because of the rotation and the flattening of the Earth at the poles, a person will weight 0.25 percent more at the poles than at the equator.

An astronaut orbiting the Earth in a satellite feels *weightless.* This phenomenon occurs because, in the frame of reference in which the astronaut is at rest, the Earth's force of attraction is balanced by *centrifugal force.* The nature of centrifugal force is explained in Figure 3-4. It is the same force that we experience in a car when it makes a sharp turn and a force seems to push us in the opposite direction.

The weight of a body on the Moon is nearly one-sixth of its weight on the Earth (see Figure 3-5) because the value of g, the acceleration due to gravity on the Moon, is one-sixth of that on the Earth. Table 3-1 shows how the weight of a person will vary on different planets.

Table 3-1

Planet	Weight as a percent of weight on Earth
Mercury	37
Venus	89
Earth	100
Mars	38
Jupiter	265
Saturn	114

*Ernst Mach has questioned the validity of this Newtonian concept that mass is an intrinsic property of a body. In his book *The Science of Mechanics,* published in 1893, Mach gives persuasive arguments against this concept. Mach's ideas have had considerable influence on the thinking of many physicists in this century, including Albert Einstein.

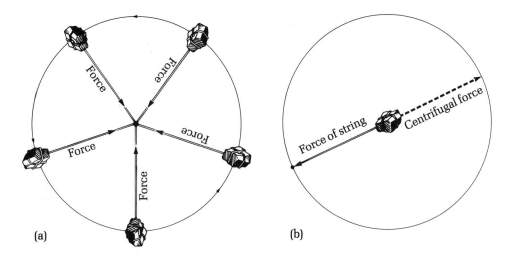

(a)   (b)

3-4. (a) A stone is whirled around in a circle at the end of a string. The string is taut; it exerts a force that pulls the stone toward the center of the circle. (b) From the point of view of the stone, the inward force of the string is balanced by an equal but opposite outward force. This is the *centrifugal force*. An astronaut in a satellite going around the Earth feels weightless; this happens because the centrifugal or outward force cancels the inward force of the Earth's gravity. (From *The Physics–Astronomy Frontier* by F. Hoyle and J. V. Narlikar. Copyright © 1980. W. H. Freeman and Company.)

3-5. A body has the same mass but different weight from location to location, depending on the force of gravity. One's weight on the Moon is one-sixth of one's weight on the Earth!

Gravity barriers

Because of one's reduced weight on the Moon, any person will be able to jump higher and throw a ball farther on the lunar surface than on the Earth. Physicists express this fact by saying that *gravity erects a taller potential barrier on the Earth than on the Moon.* Let us try to understand this statement with the help of an example.

In Figure 3-6, we see a man attempting to toss a ball into a basket located vertically above him. His job is an easy one as long as the basket is located only a few feet above his head. However, as the basket is raised to greater and greater heights, the man has to throw the ball up with greater and greater speed. In part b, we see what happens when the ball is thrown up at a speed of 40 feet per second. The graph shows how the speed of the ball is progressively *reduced* as it rises. Starting with the value of 40 feet per second from the point of ejection, the speed declines to 32 feet per second by the time the ball has risen to a height of 9 feet, to 24 feet per second at a height of 16 feet, to 8 feet per second at a height of 24 feet, and finally to zero at a height of 25 feet.

Twenty-five feet is the maximum height attained by the ball in the above example. If it did not reach the basket, it would begin to fall with increasing speeds. It would attain the same sequence of speeds at the corresponding heights that it had while rising. This relationship between height and velocity can be understood in terms of the energy–work relation discussed in Chapter 1. When the ball was rising, it was doing work *against* the force of gravity. The work done against gravity resulted in a decrease in the *kinetic energy* of the ball. By the time the ball attained the height of 25 feet, it could no longer draw upon its store of kinetic energy. It stopped rising and came to rest.

At this point, we may wonder if there has been a net loss of the ball's energy. Certainly, the ball has been progressively losing kinetic energy as it rises. This energy was lost in doing work against the gravitational pull of the Earth. Has the ball got anything to show for it when it reaches the maximum height of 25 feet? The answer is yes. The ball begins to acquire *potential energy* as it rises in just the amount that it loses kinetic energy. This potential energy

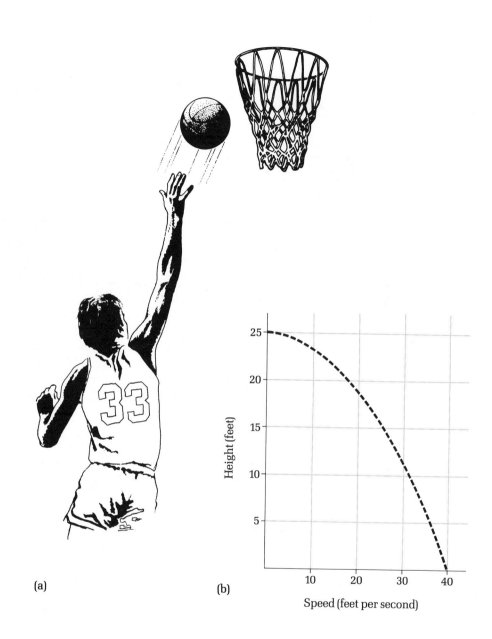

(a)   (b)

Speed (feet per second)

3-6. (a) Because Earth's gravity pulls the ball down, the thrower has to hurl it up with a certain minimum speed if the ball is to reach the basket. This minimum speed increases with the height of the basket above the ground. (b) The speed of the ball decreases as it rises. The dashed line shows this decreasing trend of speed with height.

3-7. The Bhakra Nangal hydroelectric dam in North India converts the kinetic energy of water, acquired as a result of falling from a height, into electric energy. With a height of 680 feet, it is one of the tallest straight gravity dams in the world. (Photograph courtesy of the Bhakra Beas Management Board, reproduced by permission of the Ministry of Energy, Department of Power, Government of India.)

increases in proportion to the height of the ball, being at a maximum when the ball is at 25 feet. The word "potential" indicates that the ball acquires the possibility to make gravity do work by virtue of its height. As the ball descends, its kinetic energy *increases* because gravity does work on it, while its potential energy *decreases* by an equivalent amount. In a hydroelectric dam (see Figure 3-7), potential energy is converted to another form of energy, electricity.

We can therefore arrive at the *law of conservation of energy* for the moving ball:

Kinetic energy + Potential energy = Constant

Throughout the ball's motion, whether going up or down, the total of the two forms of energy (kinetic and potential) is constant.

Having reassured ourselves that the ball's total energy is conserved, let us return to the ball thrower. His muscular power limits him to a certain speed above which he cannot throw the ball. This

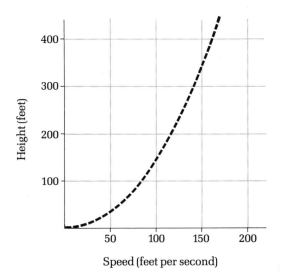

3-8. The variation of the maximum height attained with the speed of the ball. If the dashed curve were continued well beyond the range of the graph, it would tell us that, to attain a height of 4000 miles, the speed needed is about 7 miles per second. This is the escape speed for a projectile fired from the Earth's surface.

circumstance therefore puts a limit on the height to which he can throw the ball. We saw previously that this maximum height is 25 feet for a speed of 40 feet per second. A stronger thrower can of course aspire to greater heights.

Figure 3-8 shows how the maximum height varies with the thrower's beginning speed. Although the height attained increases with the speed of the throw, in each case the ball would come back to the ground if there was no basket to hold it up. No matter how strong the ball thrower is, it therefore appears that he cannot throw the ball high enough to escape the Earth altogether!

This brings us to the notion of a *gravity barrier*. To launch a spaceship to the Moon or to more distant parts of the Universe, we have to make sure that it leaves the confines of the Earth. We may imagine the spaceship to be situated at the bottom of a "gravity well." Escape from the Earth is then equivalent to climbing the wall of this well—that is, to doing sufficient work to surmount the barrier erected by the Earth's gravity. Just how tall is the Earth's gravity barrier?

Escape speed

In our previous example of the ball thrower, we have assumed the downward acceleration due to gravity to be g = 32 feet per second per second. This is the value of g on the Earth's surface, a value we first saw in connection with Galileo's experiment at the Tower of Pisa. Suppose the ball thrower has a maximum range of 100 feet above the ground. Although the strength of the Earth's attraction decreases as the ball ascends, the rate of decrease is so slow that over this height the value of g is more or less the same. So our conclusion about the relation between the maximum height attained by the ball and the initial throwing speed is not wrong; Figure 3-8 correctly describes this relationship.

However, when we extrapolate this relation to space travel, we must remember that, as we go farther and farther from the Earth, there is a significant decrease in its force of attraction. Thus the value of g at a height of 4000 miles (the radius of the Earth) is *one-quarter* of the value on the surface of the Earth. That is, g = 8 feet per second per second at this height. Hence the work done in climbing 100 feet at the altitude of 4000 miles is one-quarter of the work done in climbing the same height at the surface of the Earth. The depth of the Earth's gravity well is therefore less than what we previously estimated. The work required to lift a unit of mass out of this well is not infinite but is equal to GM/R, where M and R are, as before, the mass and radius of the Earth and G is the constant of gravitation. When we recall that on the surface of the Earth $g = GM/R^2$, this work per unit mass can be simply expressed as $g \times R$. In other words, the work done to lift a spaceship out of the Earth's gravity well is equal to climbing a height of 4000 miles *as if the value of g did not change from its surface value*.

Referring back to our previous discussion and to Figure 3-8, we therefore ask the value of the ball thrower's speed of throw for the ball to rise to a height of 4000 miles. The answer is, nearly 7 miles per second (about 11.2 kilometers per second), or 25,200 mph. This is the minimum speed that will carry a spaceship beyond the confines of the Earth's gravity. This is known as the *escape speed* for the Earth's surface.

How strong is gravity?

The escape speed gives us an indication of the strength of gravity on the surface of the Earth, and we can use this concept to compare the strength of gravity on various astronomical objects. In general, for any spherical object of mass M and radius R, the escape speed is given by the same formula that we used for the Earth,

$$V = \sqrt{\frac{2GM}{R}}.$$

On the surface of the Moon, the escape speed is only 1.5 miles per second. This means that it is considerably easier for a spaceship to leave the lunar surface than it is for the same spaceship to leave the Earth. Table 3-2 gives the escape speeds for several astronomical bodies.

Table 3-2

Astronomical object	Escape speed (miles per second)
Moon	1.5
Earth	7
Jupiter	38
Sun	400
Sirius B (a white dwarf star)	3,000
Neutron stars	about 100,000

The stronger the gravitational pull of the object, the larger is the escape speed. In Table 3-2, the neutron star has the largest escape speed, nearly two-thirds of the speed of light. A neutron star is a very compact star; its density is a million billion times that of water. A neutron star as massive as the Sun may have a radius of only 10 miles!

Is the escape speed from a neutron star the greatest speed necessary to escape from any object? Theoretical physicists tell us that, in principle, the greatest possible speed of a physical body cannot exceed the speed of light. For an object whose radius is determined by the formula

$$R = \frac{2GM}{c^2} \quad (c = \text{speed of light})$$

the escape speed equals the speed of light. In other words, even light can barely escape from the surface of such an object. Astronomers rely on electromagnetic radiation in its various forms—visible light, radio waves, X rays—to observe objects, but astronomers cannot observe an object if the escape speed is faster than the speed of light. Such an object can only be detected by its strong gravitational force.

This is our first encounter with the concept of a *black hole.*

A computer-enhanced photograph of a solar eclipse shows great streams of gas extending to the outer corona. (Los Alamos National Laboratory.)

4

Fusion Reactors in Space

4

Fusion Reactors in Space

Fusion

It is often argued that man's growing energy needs will be met if he succeeds in making fusion reactors. In a fusion reactor, energy is generated by fusing together light atomic nuclei and converting them into heavier ones. The primary fuel for such a fusion reactor on the Earth would be heavy hydrogen, whose technical name is deuterium. Through nuclear fusion, two nuclei of deuterium are brought together and converted to the heavier nucleus of helium, and in this process nuclear energy is released.

Following is the recipe for a fusion reactor. First, heat a small quantity of the fusion fuel, deuterium, above its ignition point—to a temperature of some 100 million degrees centigrade. Second, maintain this fuel in a heated condition long enough for fusion to occur. When this happens, the energy that is released exceeds the heat input, and the reactor can start functioning on its own. The third and final part of the operation involves the conversion of the excess energy into useful form, such as electricity.

The recipe appears simple, but it is not so simple to translate into reality. Worldwide research efforts over many years have yet to yield a practical solution. Nuclear fusion was achieved years ago when hydrogen bombs were made. The H-bombs are a testimony to the vast reservoir of energy in atomic nuclei that can be released through fusion. Where then is the difficulty in making fusion reactors?

The difficulty lies in achieving *controlled* nuclear fusion. We want the nuclear energy to be released steadily and not explosively as in a bomb. For this to occur, the fusion material must be properly confined and held in a stable form. This is the crux of the problem that present research is trying to solve.

The first indication that nuclear fusion can operate in a controlled fashion and generate useful energy came not from any laboratory demonstration on the Earth but from the study of stellar structure. The fact that stars shine has been noted by man since he first gazed at the night sky. Shining requires generation of light, and light is a form of energy. Where do the stars get their energy?

While looking for a clue to the stellar energy source, astronomers hit upon the concept of nuclear fusion. They demonstrated that stars can generate energy in sufficient quantity, through controlled nuclear fusion, to enable them to keep shining for millions to billions of years. How have stars managed to solve their problem of controlled fusion? As we shall now see, stars are able to do with ease what man is finding so difficult because they possess one advantage. Because of their enormous masses, stars can call upon gravity to act as the controlling agent.

To understand this role of gravity, let us use an imaginary episode from the life of Aladdin.

The genie and the Sun

The *Arabian Nights* story of Aladdin and the magic lamp ends with Aladdin living "happily ever after" with his princess and his magic lamp. Here is a postscript to the story to interest the astronomers.

One hot summer's day, Aladdin, while on a tour of the Arabian desert, suffered a sunstroke from which he took many days to recover. When he became well, he summoned the genie of the magic

4-1. The genie of the magic lamp takes apart the Sun.

lamp and issued this command: "Take the Sun apart and distribute its bits and pieces far and wide so that it is completely destroyed" (see Figure 4-1).

Now, assuming that the genie possesses boundless power and can undertake this mammoth task, just how much work is involved in the execution of Aladdin's command? Astronomers tell us that the Sun is a nearly spherical ball with a radius of nearly 700,000 kilometers and a mass* of approximately 2,000 billion billion billion kilograms.

Of course, the genie soon realized that chipping off bits and pieces from the surface of the Sun and taking them far away does require physical work. This is because each bit is attracted by the remainder according to the law of gravitation. To take any particular bit away from the rest, the genie has to work *against* this force of gravity. What is the total amount of work the genie would have to do to take the Sun completely apart and to move all its bits and pieces far away? The precise answer to this question will depend on the exact distribution of matter in the Sun. But the answer is of the order of $G \times M^2/R$, where M is the mass of the Sun and R is its radius. G is the constant of gravitation (which we encountered in Chapter 2 in the statement of Newton's law). With the values of M and R just given, this quantity of work turns out to be about 4×10^{48} ergs. We will shortly put this large quantity in proper perspective. For the time being, let us denote it by the symbol W.

Aladdin began to have second thoughts long before the genie completed the job. He realized how essential the Sun was to the inhabitants of the Earth, including himself. So, while the genie was in the process of completing the job, Aladdin issued his next command: "Put all the bits of the Sun back together." The poor genie went back to execute the command.

However, this time, to bring all the constituents of the Sun together, the genie no longer had to work against gravity. In fact, while the genie had been temporarily called away by Aladdin to issue his second command, the bits and pieces left in space by the genie had already begun to fall back together, as shown in

---

*See Chapter 3 for a discussion of how the law of gravitation enables us to measure the mass of the Sun.

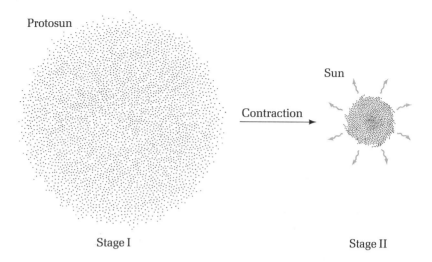

Protosun

Sun

Contraction

Stage I

Stage II

4-2. Two stages in the contraction of a gas cloud that becomes the Sun. Stage I is an early stage when the cloud is well dispersed and beginning to contract. In Stage II, the cloud has shrunk considerably because of its self-gravity. When the cloud has shrunk enough for nuclear reactions to begin, it radiates energy.

Figure 4-2. Gravity, which had been an opposing force for the first job, had now turned into an ally. And to put the Sun back together, the genie had to do no work. Instead, the amount of work *W*, which the genie had earlier expended on the first job *against* the force of gravity, would now be done by the force of gravity to put the Sun back together.

Let us now recall the concepts of work and energy outlined in Chapter 1. There we saw that, for a moving body, the work done *by* the impressed force results in an *increase* in the kinetic energy of the body. In the story of Aladdin, we note that gravity (as an impressed force) does work to bring the Sun to its present shape from an initially dispersed state. When we put these two ideas together we arrive at the *Kelvin–Helmholtz contraction hypothesis* for explaining why the Sun shines.

The Kelvin–Helmholtz hypothesis

Two distinguished physicists of the last century, Lord Kelvin (1824–1907) and Hermann von Helmholtz (1821–1894), suggested gravity as the primary source for stellar energy. Their hypothesis is called a *contraction hypothesis* because it states that the continued contraction of the Sun under its own gravity generates energy for radiation.

Consider the two states of the Sun shown in Figure 4-2. Stage I now represents an early stage in the Sun's history. In Stage I, the Sun was much bigger than its present form, shown as Stage II. If the Sun is formed through condensation of an interstellar gas cloud, Stage I represents the state when the constituents of the Sun were well spread out. This is precisely the state the Sun was brought to by the genie under Aladdin's first command! From Stage I to Stage II, the Sun contracts under its force of gravity; in other words, work is done *by* the impressed force of gravity in bringing the Sun to the present state from its primordial dispersed state.

By our rule for conversion of work to energy, this work by gravity should appear as kinetic energy. There is, however, no large-scale motion in the Sun. So where did this kinetic energy go?

If we were to examine this question carefully, we would discover that the kinetic energy has not disappeared! The Sun is in a gaseous state, and the particles of gas do move. As shown in Figure 4-3, the movement of gas particles is not systematic but *random*. Gas

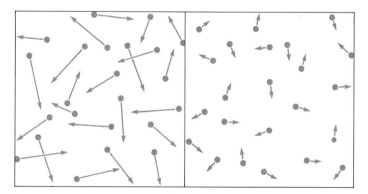

4-3. The random motion of gas particles gives rise to the property of pressure. This pressure will be exerted on the enclosure containing the gas; the gas particles bounce against the walls and are reflected. This bouncing produces force on the wall by Newton's third law of motion. The magnitude of the pressure is related to this force and it depends on the density and temperature of the gas.

particles move in all possible directions with speeds ranging from small to large. Although these motions average out, leading to no systematic large-scale motion, the gas does have kinetic energy. And this energy increases (the gas particles move faster and faster) as the Sun slowly contracts.

If there is no manifestation of this kinetic energy in the form of a visible large-scale motion, how is the energy manifested? The effect of the kinetic energy is seen through the *pressure* of the gas. As the kinetic energy rises, the gas pressure also rises.

Even on the surface of the Earth we talk of gas pressure when we describe the daily barometer reading. The statement that "the barometer reading is 30 inches" means that there is enough pressure in the atmosphere (see Figure 4-4) to support a vertical column of mercury 30 inches high. As we go up in an aircraft, the atmospheric pressure falls. At a height of 10,000 feet, it becomes low enough to make it necessary to pressurize the aircraft.

An additional effect of the change in pressure of a gas is a corresponding change in temperature. In the contracting gas cloud that eventually became the Sun, as the pressure increased, the temperature also increased. A gas at high temperature radiates light.

So, in the contraction hypothesis of Kelvin and Helmholtz, we have the following sequence of energy conversion:

Gravitational energy → Kinetic energy → Radiation energy

The Sun shines because of its gravity.

Let us now consider W, the amount of work done by the genie. W is also the energy the Sun has expended during contraction from Stage I to Stage II. How long has the contraction gone on? To calculate this time interval, we need to know the rate at which the Sun has been expending this energy through radiation. From the amount of radiation received on the Earth, astronomers calculate this rate of energy expenditure to be about 1.2×10^{41} ergs per year. If this rate has not changed substantially from past rates, then gravitational energy has kept the Sun shining for a period of about *30 million years*.

Thirty million years is a long time span by human reckoning, and our first reaction is that the Kelvin–Helmholtz hypothesis has provided a satisfactory explanation of solar luminosity. However, difficulties with this hypothesis appeared when geologists

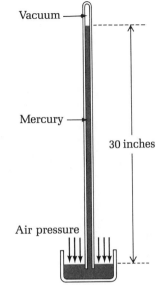

4-4. In the standard barometer, the pressure of the air (shown by arrows) is able to support the weight of a column of mercury. The height of the column gives a measure of the pressure. The part of the tube at the top is a vacuum. As the atmospheric pressure rises and falls, the column of mercury also rises and falls.

Vacuum

Mercury

30 inches

Air pressure

estimated the age of the Earth to be considerably longer than 30 million years. The present estimate of the age of the Earth is nearly *4.5 billion years!* If the present ideas about the origin of the solar system are to be trusted, the Sun and the Earth must have formed at about the same time. If the Sun is considerably older than 30 million years, we must look to some other source than gravity for an explanation of its energy reservoir.

The Sun as a fusion reactor

The mystery of the Sun's energy reservoir remained unsolved until the third decade of the present century. By then, astronomers had begun to have clearer ideas about the internal constitution of the Sun and other stars. The English astronomer Sir Arthur Eddington was able to express these ideas in the form of *four equations of stellar structure.*

These equations essentially summarize the following information. The first equation is called the *equation of hydrostatic equilibrium*

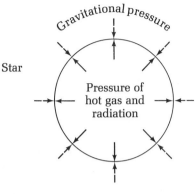

4-5. The opposing forces of gravity (inward dashed arrows) and pressure (outward solid arrows) are shown acting on any spherical surface inside and concentric with the surface of a star. For the hydrostatic equilibrium of the star, these forces must be in exact balance.

(see Figure 4-5). It describes how the Sun (or a star) is held in equilibrium under the opposing forces of gravity and internal pressures. The internal pressure in a star partly arises from the hot gas in its interior and partly from the radiation emitted by the hot gas. Gravity has the tendency to shrink the Sun, whereas internal pressures tend to expand it. The second equation describes how the mass of the Sun is related to its density. The third equation, known as the *equation of state,* connects the pressure to temperature and density. The overall effect of these equations is to generate a model of the Sun as a sphere of gas with a high temperature in the center that progressively decreases outward. The fourth equation describes how the radiation generated in the hot inner regions is progressively absorbed as it moves outward. Because of the absorption, only a small fraction of the inner radiation can escape from the surface.

With these equations, Eddington was able to show that a viable model of the Sun has a surface temperature of about 5500° centigrade (astronomers had previously estimated this temperature by an analysis of the Sun's radiation) and a central temperature higher than 10 million degrees centigrade.

The missing piece of information at this stage was the mysterious energy reservoir of the Sun. Here, Eddington made a prophetic suggestion. He argued that the central temperature in the Sun was high enough to release nuclear energy in sufficient quantities to provide for the Sun's luminosity.

The atomic physicists disagreed. They felt that stellar interiors are not hot enough to trigger release of nuclear energy. To such

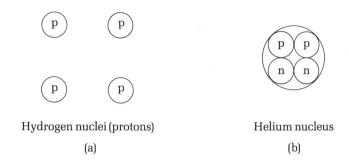

Hydrogen nuclei (protons)   Helium nucleus

(a)   (b)

4-6. In (a) we have four hydrogen nuclei, which are in fact
four positively charged protons (p). In (b) we see the helium
nucleus with two protons and two electrically neutral particles,
neutrons (n). In a star, the material in (a) is transformed to the
material in (b). In this process some energy is released.

critics, Eddington gave this spirited reply*: "We do not argue
with the critic who urges that the stars are not hot enough for this
process; we tell him to go and find a hotter place." Hell hath no
fury like a theoretician scorned!

In the 1920s, nuclear physics was a new subject, and neither
Eddington nor his critics had enough knowledge to pursue the
argument further. Eventually Eddington was proved right—the
temperatures in the central regions of stars *are* high enough to
sustain nuclear reactions with the fusion of light atoms.

With our present knowledge of the nucleus, it is possible to
understand why the controversy arose in the first place and how it
was subsequently resolved. In Figure 4-6a, we see four separate
hydrogen (H) nuclei, which are in fact the positively charged
elementary particles called *protons*. In part b, we see the nucleus
of a helium atom (He). This has two protons and two *neutrons*.
Neutrons are uncharged particles. In a fusion reaction, the four H
nuclei are brought together and converted to a He nucleus:

$$4H \rightarrow He + 2e^+ + 2\nu + Energy$$

*See A. S. Eddington, *Internal Constitution of Stars,* Cambridge University
Press, 1926, p. 301.

This symbolic way of writing the reaction tells us that the products of the reaction are a helium nucleus, two positrons (e^+), two neutrinos (ν), and release of energy. *Positrons*, the antimatter counterparts of *electrons*, have the same mass as electrons but are positively charged. Indeed, if we were to require that the total electrical charge remain unchanged in a fusion reaction, then it follows that, besides the He nucleus, two units of positive charge must be carried by other reaction products. Positrons do this job.

The release of energy in the above fusion reaction takes place for the following reason. The masses of the four participating H nuclei taken together exceed the sum of the masses of the reaction products (the He nucleus and the other four light particles) by a small amount. Einstein's special theory of relativity states that loss of mass in any process in nature is balanced by a corresponding gain in energy. This energy is related to the loss of mass by the famous formula $E = Mc^2$.

In the fusion reaction leading to the formation of the He nucleus, the mass that is lost is equivalent to the energy of 26.72 MeV*. To put it differently, of the hydrogen converted to helium, the fraction of mass that is converted to energy is 0.7%. This is the energy reservoir that man is trying to tap through his attempts to build a fusion reactor.

The reaction in the fusion reactor differs somewhat from fusion in the Sun. In the fusion reactor on Earth, the primary fuel is *heavy* hydrogen. Its nucleus has a neutron as well as a proton. *Two* such nuclei have to be fused to yield a He nucleus and radiant energy.

The atomic physicists of the 1920s objected to Eddington's hypothesis because of the difficulty of bringing four H nuclei together. Since protons are positively charged, they repel each other according to the electrostatic law that *like charges repel*. How then could these like charges be brought together? The difficulty looked insurmountable in the 1920s but was resolved in the next decade with the discovery of the attractive nature of the strong nuclear force. Notice that the He nucleus of Figure 4-6b has two protons

---

*MeV, or 1 million electron volts, is an atomic unit of energy. One kilowatt hour, the unit commonly used for measuring electricity consumption, equals about 2×10^{19} MeV.

held together. How can this happen if two like charges repel each other? The answer is that, within the nucleus, a new force, much stronger than the electrostatic force of repulsion, acts in such a way as to hold the four particles (the two neutrons and two protons) together. This strong nuclear force acts on neutrons as well as on protons, but it acts over a very short range. If protons can bombard each other with sufficient speeds, they can come close enough to feel the effect of the strong nuclear force. In a high-temperature hydrogen gas, the H nuclei would have large random motions and thus would occasionally surmount the electrostatic repulsion and come close enough to be fused by the strong nuclear force. The temperatures in the centers of stars, which range from 10 to 40 million degrees centrigrade, are high enough to endow the H nuclei with great enough speeds to bring them together, as Eddington had argued.

Gravity as the controlling agent

The modern theory of stellar structure is based on the four equations set up by Eddington, together with a fifth equation describing the rate of energy generation in the fusion reactions in the central core of the star. In 1938, Hans Bethe solved the fifth equation and constructed a complete model of the Sun.

The crucial role of gravity in these equations cannot be over-emphasized. Huge pressures are needed inside the Sun in order to balance the attractive force of gravity and prevent total contraction. These pressures are connected with high temperatures and densities. A contracting cloud of interstellar gas becomes a star when the temperature in the core reaches a high enough value to trigger nuclear reactions.

In man's attempts to produce high temperatures suitable for triggering nuclear reactions, the controlling effect of gravity is absent. In the deep interior of the Sun, gravity confines the gas undergoing explosive nuclear energy generation. On Earth, man has to look for other means, such as a magnetic field, to confine the hot gas. These attempts are still a long way from success.

Let us now make a thought experiment to further explore the effect of gravity on stars. Suppose we connect a hot star to a cold star by a conducting wire. We know that heat flows from a hot body to a cold body, and accordingly in our thought experiment heat will flow from the hot star to the cold star.

Nevertheless, there is a surprise in store for us! Under normal circumstances, when heat flows from the hot to the cold body, the temperature of the hot body is lowered while that of the cold body rises. In our example, when heat passes from the hot star, its internal pressures drop and its equilibrium is disturbed. Because its pressures are no longer strong enough, the star contracts under its force of gravity (see Figure 4-5). As the star contracts, its gas heats—thus the hot star becomes hotter! What happens to its cold companion? As it receives heat, its pressures rise and its equilibrium is also disturbed. This star *expands* as its internal pressures become stronger than its force of gravity. As the star expands, its gas cools further. So the cold star becomes colder!

Strange though this behavior is, something like it does occur during the evolution of a star. We have already seen that, in the central core of a star like the Sun, the temperature is high enough to sustain the fusion of hydrogen into helium. What will happen when the hydrogen in the core is exhausted? The fusion reactor will be temporarily switched off for want of fuel. This will lead to a drop in the heat production and in the pressure in the core. The core therefore shrinks and heats up. As its temperature rises and reaches, say, 100 million degrees centigrade, the reactor comes alive again. However, now the fuel to be burnt is not hydrogen but helium. At this temperature, three helium nuclei can be fused to form a carbon nucleus, as shown in Figure 4-7. Meanwhile, to preserve overall equilibrium, the star's outer envelope *expands*. The star becomes a *giant star*. The expansion of the envelope leads to cooling, so that the surface temperature of the star drops. Whereas the Sun has a surface temperature of about 5500°C, a giant star can have a surface temperature as low as 3500°C. The color of a giant star is therefore closer to red, in contrast to the predominantly yellow color of the Sun (see Figure 4-8).

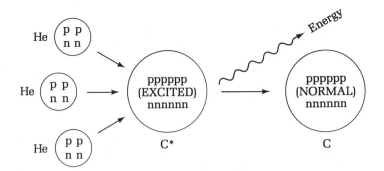

4-7. The diagram shows that three helium nuclei, each containing two protons and two neutrons, are fused into a nucleus of carbon. Fred Hoyle has pointed out that this nucleus is in an *excited* state and it decays into a normal carbon nucleus with the emission of energy. This nuclear reaction occurs in later stages of stellar evolution when a star's hydrogen fuel has been fused into helium nuclei. Helium fuses at a higher temperature than hydrogen.

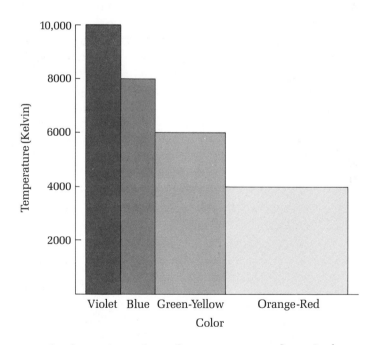

4-8. This figure shows the surface temperatures of stars in degrees Kelvin, which correspond roughly to the different colors in the visible part of the spectrum. The temperature increases as we go from red to violet. Thus a star with a temperature of around 8000 K will appear blue, whereas a star with a temperature of around 4000 K will appear mainly orange-red.

Stellar catastrophes

During the course of stellar evolution, this process of expansion and contraction goes on and on. While fuel is available, the star burns it. When the fuel is exhausted, the core contracts and heats up until it attains a high enough temperature to activate the fusion process with new fuel. In this sequence, a series of heavier and heavier nuclei are built up:

$$\text{Helium} \rightarrow \text{Carbon} \rightarrow \text{Oxygen} \rightarrow \text{Neon} \rightarrow \text{Silicon} \rightarrow \text{Iron}$$

At each stage in this process, the outer envelope of the star expands further to maintain equilibrium. The giant gets larger and larger. Nuclear physics tells us, however, that the process of fusion cannot just go on and on. In fact, it stops at the iron group of nuclei. Any further fusion of particles to the iron nucleus does not produce more energy. At this stage, the core temperature has reached around *10 billion degrees* centigrade, and the star encounters a catastrophic situation. Gravity, which was so far able to exercise a restraining influence on the hot star, can no longer do so. Instabilities develop in the star, which results in the blowing away of its outer envelope (see Figure 4-9).

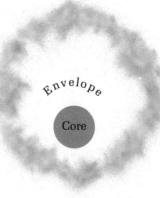

4-9. A supernova explosion occurs when the outer extended envelope of a massive star is blown off due to internal instabilities.

This stellar catastrophe is seen as a *supernova* explosion. The photograph of the Crab nebula shown in Figure 4-10 is the most visually spectacular example of such a supernova explosion. The ancient records of Chinese and Japanese astronomers tell us that the explosion itself must have been observed from the Earth on July 4, 1054. The photograph shows the expanding outer envelope now, some nine centuries after the explosion.

The fallout of such an explosion takes the form of atomic nuclei (which have been fused inside the star), electrons, neutrinos, and radiation. The nuclei appear as showers of *cosmic rays*, which travel long distances in our Galaxy. It would indeed be catastrophic for us on the Earth if a supernova explosion occurred within a distance of, say, 100 light-years. The high-energy cosmic rays resulting from such an explosion would play havoc with the Earth's atmosphere. For example, they could strip it of all its protective

4-10. Photograph of the Crab nebula. The bright object is believed to be the debris of a supernova explosion. (Courtesy of Mt. Wilson and Las Campanas Observatories, Carnegie Institution of Washington.)

layer of ozone, thus exposing life on Earth to the devastating ultra-violet rays from the Sun. Fortunately, a supernova explosion is not very common. Throughout the Galaxy, the frequency of such explosions may be once in 100 to 300 years. So the chance of such an explosion occurring in our neighborhood, up to a distance of 100 light-years, is as small as one part in a million per thousand years.

Destructive though a supernova explosion appears to be, there is evidence that such an event may itself trigger star formation in a nearby gas cloud. The composition of our solar system suggests that its birth may have been triggered by such a supernova explosion. The shock waves generated from such explosions, impinging on an interstellar gas cloud, can set off its contraction. The Sun and the planets may have condensed out of such a contracting gas cloud. Thus stellar catastrophes can in this way play a constructive role as well as a destructive role.

What about the remnant of this explosion? What is left behind after the envelope has been cast off? We will return to this subject in Chapter 7.

The two quasars 0957 + 561 A,B (aligned vertically in the center of the photograph) are not two separate quasars but a single object whose light has been split into two images by the gravitational field of a galaxy located between the quasar and our solar system. The intervening galaxy, which is located just above the lower quasar image, actually forms three images of the quasar, one deflected to the north of the actual position and two deflected south so that they overlap each other. (Courtesy of David H. Roberts.)

5

Living in Curved Spacetime

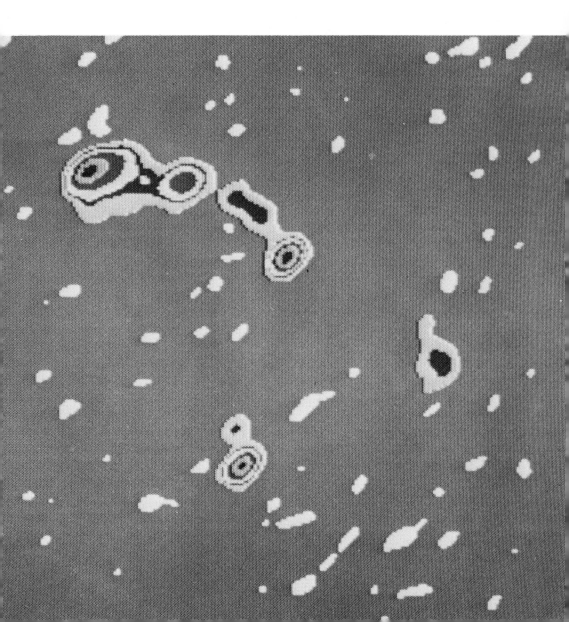

5

Living in Curved Spacetime

Is Newton's law perfect?

We left Chapter 2 with the impression that Newton's law of gravitation gave a successful account of the diverse natural phenomena in which gravity is believed to play a leading role. Not only is this law able to account for motions of such celestial bodies as planets, comets, and satellites, it also helps us in understanding the complex problem of the structure and evolution of the Sun and other stars. Modern scientists use the same law in *determining* the rocket thrusts, spacecraft trajectories, and the timing of space encounters. That a good scientific law should be basically simple but universal in application is epitomized in Newton's law of gravitation. What more could one ask for?

Yet science by nature is perfectionist. The laws and theories of science are accepted as long as they are able to fulfill its primary purpose of explaining natural phenomena. Any law of science with a history of past successes is inevitably discarded if it fails in even one particular instance. To the scientist, such an event brings mixed feelings. Disappointment and confusion that an old, well-established idea has to be given up or modified are coupled with excitement and expectation that nature is about to reveal a new mystery.

Newton's law of gravitation was no exception to this rule. By the beginning of the present century, cracks were beginning to appear

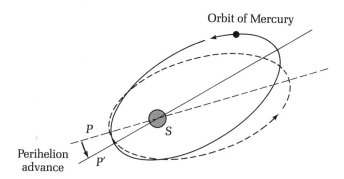

5-1. The precession of the perihelion of the planet Mercury. (From *New Horizons in Astonomy* by J. C. Brandt and S. P. Moran. Copyright © 1972. W. H. Freeman and Company.)

in the impressive facade of physics erected on the Newtonian ideas of motion and gravitation. The cracks were both conceptual as well as observational. It would take us too long to describe all the different issues involved, so we will limit ourselves to one example of each, taking the observational discrepancy first.

The strange behavior of Mercury

The orbits of planets are supposed to be ellipses. This was Kepler's conclusion after a careful analysis of the data, a conclusion that was subsequently "proved" by Newton on the basis of his laws of motion and gravitation.

However, observations extending over several decades after 1764 began to reveal a minor discrepancy in the orbit of the planet Mercury. Of all the planets of the solar system, Mercury is the closest to the Sun and has the most eccentric orbit. It takes nearly 88 days to complete one orbit around the Sun. Thus, if Mercury is at the closest point P to the Sun on a certain day, it is expected to return to that point after 88 days. This expected behavior is shown in Figure 2-5.

Mercury, however, behaves anomalously in this respect. After completing the orbit, it does not return to the same spot. As shown in Figure 5-1, the point of closest approach to the Sun will have moved from P to P'.

The point of closest approach is called *perihelion*. The situation illustrated in Figure 5-1 implies that the perihelion of Mercury has advanced; and, as Mercury keeps going around the Sun, its perihelion will advance in the direction of the arrow shown in Figure 5-1.

The rate of advance is very slow. The line joining the Sun S to the perihelion moves around in space so slowly that over 100 years it turns through an angle of only 575 seconds of arc. To put this in proper perspective, we should recall that an arc-minute is one-sixtieth of a degree, and an arc-second is one sixtieth of an arc-minute. Thus, in 100 years, the perihelion advance is only 9.58 arc-minutes, or about 0.159 degree.

Minor though this discrepancy is, its existence bothered the scientists. Recall that in Chapter 2 we encountered the case of a discrepancy in the orbit of the planet Uranus. There the apparent discrepancy was resolved when it was found that another (hitherto undiscovered) planet was perturbing the orbit of Uranus. Could the anomalous behavior of Mercury be caused by a perturbing influence of other planets?

Calculations showed the answer to this questions to be "almost, but not completely." Of the total angle of 575 arc-seconds, a major part of about 532 arc-seconds is indeed due to the perturbing effect of other planets of the solar system. The residual effect of a perihelion advance rate of some 43 arc-seconds per century remained unaccounted for in the Newtonian framework. Notice that the unexplained effect is less than 8% of the total effect and in absolute terms is extremely small. Nevertheless, to the perfectionist view of science, the surviving discrepancy did cast doubts on the validity of Newton's law of gravitation.

From Newton to Einstein

Conceptual difficulties rather than observational discrepancies led Albert Einstein (1879–1955) to cast a critical look at the Newtonian law of gravitation. The most bothersome aspect of the law of gravitation was its concept of *instantaneous action at a distance*.

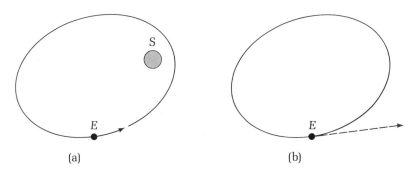

5-2. If the Sun were to disappear by magic, the Earth, freed from the Sun's gravity, would take off in the tangential direction to its elliptical orbit. (From *The Physics–Astronomy Frontier* by F. Hoyle and J. V. Narlikar. Copyright © 1980. W. H. Freeman and Company.)

The Sun and the Earth attract each other, according to this law, by a force that acts across the vast distance separating them—a distance of some 90 million miles. Not only is this force supposedly acting at great distance, it supposedly also acts *instantaneously*.

This instantaneous nature of gravity is illustrated by a thought experiment, for which once again we call upon Aladdin and his genie. Recall the incident in the previous chapter where Aladdin commanded the genie to take apart the Sun. Suppose instead that he had told the genie to annihilate the Sun at once by magic and without leaving a trace. Now, in the real world governed by the laws of physics, there are limitations on what the genie can do. For example, the law of conservation of matter and energy tells us that even the most powerful genie in the world cannot annihilate the Sun "without leaving a trace." If matter is destroyed, an equivalent amount of energy (given by the Einstein relation $E = Mc^2$ encountered in Chapter 4) must appear. But in the world of the *Arabian Nights* everything is permitted, and so let us continue our speculation on what would happen if Aladdin's command were carried out.

As shown in Figure 5-2, the moment the Sun disappears, the Earth will feel itself free from the gravitational pull toward the Sun. It will consequently take off in the direction tangential to its regular elliptical orbit. To the people on the Earth living on the part facing the Sun, the Sun's disappearance will be a visible fact (the day turning into night) some eight minutes *after* the Sun is destroyed, because light takes nearly eight minutes to travel from the Sun to the Earth. Thus, here we have a situation where an observable effect (the disappearance of the Sun) is communicated to the Earth by gravity faster than it is communicated by light.

Such a situation is contrary to the *special theory of relativity* developed by Einstein in 1905. This theory places an upper limit on the speed with which an observable effect can propagate from one point in space to another. *This upper limit is represented by the speed of light.* Thus the notion of instantaneous propagation of gravitational pull across vast distances is inconsistent with the basic tenet of special relativity. This is why, having developed the theory of special relativity, Einstein was compelled to revise the law of gravitation.

At this stage, the reader may ask "What is this special theory of relativity? Why is it so important that to conform with it we must change the law of gravitation so well established over two centuries?" In this book, we are concerned mainly with the subject of gravity and hence cannot devote much space to these interesting and important questions. The following brief discussion, while sufficient for our purpose, hardly does justice to the important place that special relativity occupies in modern physics. The revolution brought about by this theory in modern physics is comparable to the revolution brought about by Galileo to the medieval physics dominated by Aristotle.

Special relativity questioned the validity of the fundamental concepts of *absolute space* and *absolute time*. In Figure 5-3a, we see two points, P, Q in space. If we are asked to find the distance between P and Q, we would place a ruler along the line PQ and measure the length of the segment. In Figure 5-3b, we show two instants of time at A and B. To measure the period elapsed between A and B, we use a clock. The number of ticks of the clock between A and B gives us the measure of the time interval between A and B. Now, our intuitive feeling is that these measurements have an *absolute* character, that is, they do not depend on the observer.

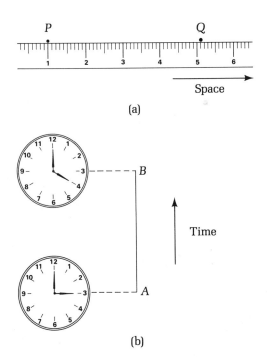

5-3. The measurements of spatial displacements (a) and time intervals (b) in Newtonian/Galilean dynamics.

In particular, if two observers moving with different speeds make these measurements, they will come up with the same answer. This is the intuitive feeling on which the physics of Galileo and Newton was based. This was the notion that Einstein challenged.

Of course, concepts in science may have their origin in the intuitions of bright scientists, but their ultimate validity rests on physical experiments. Recall from Chapter 2 that Galileo himself emphasized the role of experiments in his reasoning.

The concepts of absolute space and absolute time seemed to rest on solid foundations. Yet, toward the end of the nineteenth century, with improvement in the accuracy of laboratory experiments, cracks began to appear in these foundations. One experiment that played a very important role in this context was the experiment of

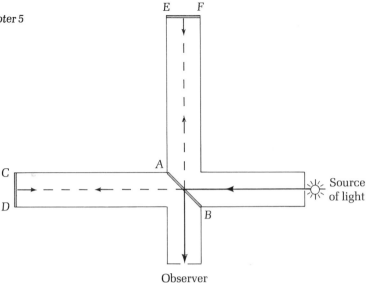

Observer

5-4. In the Michelson–Morley experiment, a ray of light falls on a partially reflecting and partially transmitting mirror AB in Michelson's interferometer. The reflected part goes up and is reflected by the mirror EF. The transmitted part goes in the original direction and is reflected by the mirror CD. The two parts recombine and are seen by the observer. If we take into account the fact that light is a wave, the result of the above recombination depends critically on the phases of the two split waves. In the extreme case of the crests of the two waves falling together, the total light doubles, while in the opposite extreme the crest of one wave can cancel the trough of another. In general, a series of dark and white fringes are seen by the observer. This interference of two waves depends on how far each wave has traveled as well as on the speed of light. Since the two arms of the interferometer are equal in length, the shifts in the interference fringes could be used to detect minute changes in the speed of light. Michelson and Morley used this technique to measure the expected difference of light travel time in the north–south and east–west directions. They failed to find any difference.

Michelson and Morley in 1887. The experiment, which is described in Figure 5-4, showed that, in an interferometer, the time taken by light to make a return journey in the east–west direction is the same as the time taken for a return journey of the same length in the north–south direction. The experiment was accurate enough to detect the effect of the (west-to-east) motion of the Earth's surface as it rotates relative to "absolute space." The surprising result that no effect was detected led to considerable speculation among the leading physicists of the nineteenth century.

In 1905, Einstein correctly resolved the implications of the Michelson–Morley experiment. Einstein had set out to reexamine the concepts of absolute space and absolute time in terms of Clerk Maxwell's laws of electricity and magnetism, formulated in the 1860s. He had come to the conclusion that, in order to preserve the inherent symmetries of Maxwell's laws, it was necessary to revise the Newtonian concepts of absolute space and absolute time. Einstein interpreted the Michelson–Morley experiment in the most direct and straightforward terms—that the speed of light was unaffected by the motion of the Earth. In fact, this was a special case of Einstein's more general conclusion that the speed of light is the same for all observers moving relative to each other.

The first law of motion, originally discovered by Galileo, singles out a special class of observers—those that move with a uniform velocity, that is, those observers on which no force acts. Such observers are called *inertial observers*; any two inertial observers move with uniform velocity relative to each other. According to Einstein, the speed of light will be the same as measured by them both. This conclusion is a special case of the more general statement that the basic laws of physics are the same for all inertial observers.

If we observe a passing express train from the platform of a railroad station, it appears to flash past. If we observe the same train from a fast-moving car on a parallel highway going in the same direction, the train does not appear to move as fast. This is because the velocity of the train relative to the car is much less than it is relative to the stationary platform.

Contrast this result with Einstein's conclusion that the speed of light is the same with respect to all moving inertial observers and you begin to see why such a conclusion is against our intuition. But, as the Michelson–Morley experiment so conclusively demonstrated, nature does not always behave according to our intuition!

Faced with this startling conclusion, Einstein had to revise the concepts of measurements of spatial distance and time intervals. The notions of absolute space and absolute time, as illustrated in Figure 5-3, were inconsistent with the constancy of the speed of light. Instead, space and time measurements became *dependent on* the inertial observer making the measurements. Two observers in relative motion making measurements of the type shown in Figure 5-3 *will not get the same answers.* The rules connecting their measurements are known as the *Lorentz transformation,* after the physicist Hendrik Anton Lorentz.* Thus, in special relativity, the Galilean concepts of absolute space and absolute time were replaced by a unified concept of space and time in which neither space nor time by itself had an absolute status. The laws of motion written according to the new rules of the Lorentz transformation implied that several changes needed to be made in the concepts of Newtonian dynamics. In particular, the new rules implied that the speed of light represents an upper limit on the speed at which any material particle or physical information can be transmitted.

Against this background, in the years 1905 to 1915, Einstein looked for a theory of gravitation that possessed all the successful features of the Newtonian law of gravitation and yet was free from the conflicts between the Newtonian law and the theory of special relativity.

The General Theory of Relativity

One of the difficulties that threatened the coexistence of the Newtonian law of gravitation with special relativity has already been mentioned. Special relativity attached a special significance to the speed of light as the upper limit that cannot be exceeded by any physical interaction, whereas gravity (according to Newton) seemed to be operating instantaneously across vast distances. Another difficulty related to the definition of inertial observers. An inertial observer is one on whom no force acts. Can we actually pinpoint any observer or any physical object that is free from all

---

*Lorentz had given these rules in connection with another theory in his own attempts to explain the result of the Michelson–Morley experiment. Einstein found these rules applicable to his special relativity, although in a different context.

5-5. Einstein managed to fit a square peg into a round hole by modifying both the peg and the hole! His general theory of relativity resolved conflicts between Newton's theory of gravity and the theory of special relativity.

forces? In our discussion of the first law of motion, we encountered friction as a force that retards motion. However, under idealized conditions, we can achieve situations where the friction is made very, very small. A man attempting to walk on an ice rink knows how difficult it is to proceed with very little friction. A projectile fired in a vacuum chamber encounters no resistance from air. In these cases, however, there is another force that has been ignored, the force of the Earth's gravity. Gravity acts on all material objects and can never be eliminated as a force. Even if we go far, far away from the Earth, we still have other objects in the Universe to contend with.

In short, there is nowhere in the Universe that we can go to eliminate gravity as a force, and so our definition of the inertial observer seems unrealizable in practice. Since the inertial observers form the starting point of special relativity, it looks as if the theory is based on unrealistic foundations. Thus, not only does special relativity make the Newtonian law of gravitation inconsistent, but its own foundations appear threatened by the existence of gravity.

Einstein discovered an ingenious way out of these difficulties by proposing an entirely new approach to the phenomenon of gravity. His theory of gravitation, called the *General Theory of Relativity*, is not a patchwork repair job on Newtonian gravitation and special relativity; rather, it is a radically new attempt at understanding the

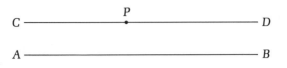

5-6. In Euclid's geometry, the parallel axiom states that through P one and only one straight line can be drawn parallel to AB. The line CD is such a line.

nature of gravity. Let us try to understand the reasoning that leads to this remarkable theory.

We have already seen how all-pervasive gravity is and the futility of trying to avoid its existence anywhere in the Universe. Einstein took this property of gravity as evidence that it is intimately linked with another all-pervasive entity around us, *spacetime*. For the linking agent between gravity and spacetime, Einstein proposed *geometry*.

Geometry is a branch of mathematics that is devoted to the study of measurements of angles and lengths of various shapes and figures. The rules of geometry were first systematically stated by the Greek mathematician Euclid (ca. 300 BC). Starting from certain apparently reasonable axioms, Euclid developed theorems about triangles, squares, circles, and other figures. For a long time, mathematicians believed Euclid's axioms to be absolutely true and incontrovertible, and Euclid's geometry acquired a unique status as *the* system describing any measurements made in space.

However, axioms are assumptions whose validity cannot be proved. Any set of axioms that are logically self-consistent can form an independent branch of mathematics. Euclid's geometry is merely one example of many possible self-consistent sets of axioms.

For example, one of Euclid's axioms relates to the existence of parallel lines. In Figure 5-6, we see a straight line AB and a point P outside it. Euclid's axiom asserts that one and only one line can be drawn through P parallel to AB. The line CD shown in Figure 5-6 is parallel to AB—that is, AB and CD will not meet even if they are extended in both directions. To our intuition, this assumption looks reasonable and *true*. But it is nevertheless an assumption.

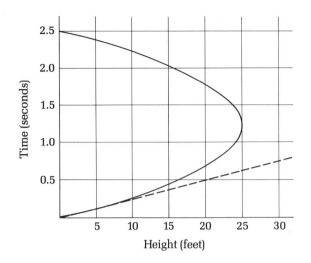

5-7. The world line of the ball is shown by the solid curve. The dashed straight line shows what this world line would have been in the absence of gravity.

It cannot be proved on the basis of other axioms. Indeed, many mathematicians mistakenly thought that this axiom *could be proved*, and some even offered what eventually turned out to be erroneous proofs. Only in the last century was it finally realized that Euclid's parallel axiom is an assumption that cannot be proved. Moreover, mathematicians were able to show that, by altering Euclid's parallel axiom, new geometries could be created that were internally self-consistent. This axiom could be altered in two ways—either by asserting that through *P no line* can be drawn parallel to *AB* or that through *P more than one line* can be drawn parallel to *AB*. The work of Lobatchevsky (1793–1856), Bolyai (1775–1856), Gauss (1777–1855), Riemann (1826–1866), and others led to whole new geometries based on the alteration in the parallel axiom. These geometries came to be known as *non-Euclidean geometries.*

Let us now return to Einstein and his attempt to link gravity to spacetime. To illustrate Einstein's reasoning, we will revive our example of the ball thrower in Chapter 3. In Figure 5-7, we illustrate the trajectory of the ball in a spacetime diagram. Recall that the ball was thrown up with an initial speed of 40 feet per second, reached a height of 25 feet, and then fell back. The trajectory shown as a continuous curve in Figure 5-7 tells us where the ball is to be found at any given instant during its flight. Such a trajectory is called the *world line* of the ball. The world line in Figure 5-7 is a curve known as a *parabola*.

Let us now speculate what would have happened to the ball *if there were no gravity*. By the first law of motion, the ball would have continued moving with the same speed (40 feet per second) in the upward direction. Its world line then would be shown by the dashed straight line.

A Newtonian scientist would say that, in the absence of gravity, the straight line describes the ball's trajectory, whereas with gravity the straight line is bent to the parabolic shape. To this statement, Einstein would have replied that there is no such thing as the absence of gravity, and hence the dashed line of Figure 5-7 has no real status. The only line of physical significance is the curved line. How should one interpret the curved line by itself?

Einstein's reply to this question was that this line in fact describes the motion of the ball under no forces but in a spacetime whose geometry is changed by gravity. Because of the Earth's gravity, the spacetime has acquired a *non-Euclidean geometry*, and in that geometry the ball's trajectory is straight.

This may appear crazy. You might be tempted to say that the continuous curve in Figure 5-7 is obviously not straight, but let us ponder Einstein's reply. How do we define a straight line? It is the line of shortest distance between two points. The shortest distance between two points is different in a non-Euclidean geometry than in Euclid's geometry. So what may be a straight line in one geometry may not be a straight line in the other, and vice versa. Thus, provided we know the nature of the non-Euclidean geometry in the spacetime around the Earth, we can test the veracity of Einstein's claim that the continuous line of Figure 5-7 is straight. How do we find out the details of this geometry?

Einstein's general theory of relativity gives us the method for determining this geometry. Einstein's equation tells us, in principle, how the geometry of spacetime is related to the distribution of matter and energy. I have said "in principle" because these equations are very complex and *in practice* these equations have

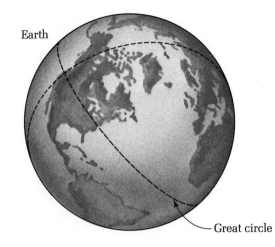

Earth

Great circle

5-8. Straight lines on the surface of the Earth are arcs of great circles, that is, circles whose planes pass through the center of the Earth. Notice that any two great circles meet on the Earth's surface.

been exactly solved in very few cases. Thus the question of determining the spacetime geometry is a difficult one. Later we will discuss one solution that has proved immensely useful in testing Einstein's theory.

Non-Euclidean geometries

Before we consider observable implications of Einstein's theory, let us become familiar with the somewhat strange concept of non-Euclidean geometries. To begin, let us consider the surface of the Earth, idealized to be a perfect sphere. Imagine that the inhabitants of the Earth are flat, two-dimensional creatures on this surface, with no appreciation of the third dimension of height. How will these creatures determine straight lines? Applying the prescription of shortest distance, two creatures will stretch a piece of string between two points so that it is taut and lies on the surface of the Earth (see Figure 5-8). The resulting line is a segment of the great circle through the two points. This is precisely the prescription that pilots would use to determine the shortest flight path between

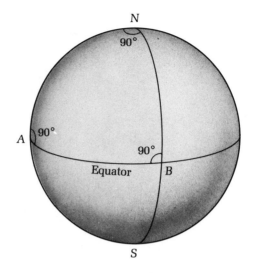

5-9. The triangle *NAB* has three right angles.

two airports. And this flight path will be different from the one gotten by using a ruler to draw a straight line on a flat map on paper. This illustrates once again our earlier remark that what constitutes a straight line depends very much on the basic rules of geometry. The geometry on the flat map is Euclid's, whereas that on the Earth is non-Euclidean. For determining the path of the aircraft, it is the non-Euclidean geometry that is relevant.

There is another way in which we may think of a line as being straight. As we move along a curve, if we find that our direction of motion does not change, then we say that we are moving along a straight line. In Euclid's geometry, the straight line defined by the "shortest distance" criterion also has this property. In a non-Euclidean geometry, these two criteria for determining whether a line is straight may differ from each other. However, in the non-Euclidean geometry chosen by Einstein (known as *Riemannian geometry*), these two criteria give the same answer.

Is there a simple method by which we can test whether the geometry on the surface of the Earth is non-Euclidean? Refer to Figure 5-8 and notice that *any two straight lines on the surface of the Earth intersect.* Thus Euclid's parallel axiom cannot be satisfied—no straight line can be drawn through *P* parallel to *AB*.

There are other ways we can test the non-Euclidean character of the Earth's surface geometry. In Figure 5-9, we have a triangle

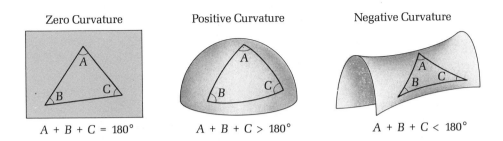

Zero Curvature   Positive Curvature   Negative Curvature

$A + B + C = 180°$ $A + B + C > 180°$ $A + B + C < 180°$

5-10. Examples of two-dimensional surfaces of zero, positive, and negative curvature. The three angles *A,B,C* of a triangle sum differently in the three cases.

drawn on the surface of the Earth that describes a journey undertaken by a flat creature starting from the North Pole *N*. The journey proceeds in a straight line—which happens to be a meridian line—up to the equator. There the creature turns left and describes the segment *AB* equal to a quarter of the equatorial circumference of the Earth. At *B* he turns left again and returns to *N* via the meridian *BN*. At *N* he finds that his direction of return is at right angles to his starting direction. Now this triangle *NAB* has a right angle at each of its three vertices, whereas the three angles of a Euclidean triangle add up to only 180 degrees. This example brings us to the notion of a *curved space*.

In the language of geometry, the two-dimensional space of the Earth's surface is *curved* and has *positive curvature*. The two-dimensional space on a flat piece of paper is *flat* or of *zero curvature*. The space on the surface of a saddle is also *curved* with *negative curvature*. These curved surfaces are illustrated in Figure 5-10. Notice that, for a triangle drawn on a space of negative curvature, its three angles add up to *less than* 180 degrees. Another way to decide whether a two-dimensional surface has a positive, negative, or zero curvature is to stretch a piece of paper on it and attempt to cover it. If the paper covers the surface exactly, the surface has zero curvature; if the paper is wrinkled, the surface has positive curvature; if the paper is torn, the surface has negative curvature. Try these criteria on various curved surfaces.

Do we live in curved spacetime?

We now return to Einstein's interpretation of gravitation and ask whether there is any observational evidence that we live in a curved spacetime.

In 1916, soon after Einstein proposed his theory of relativity, Karl Schwarzschild (1873–1916) solved Einstein's equations to find out how the geometry of spacetime behaves if there is a massive spherical object in it. The *Schwarzschild solution* is the analogue of the Newtonian solution to the problem of how a spherical mass attracts other bodies gravitationally. This solution can therefore be used to determine, for example, how planets move around the Sun.

Recall Einstein's interpretation of the first law of motion in the presence of gravity. A planet's orbit around the Sun will be determined, according to this interpretation, by the criterion that it is a "straight line" in the spacetime whose geometry is given by Schwarzschild's solution. And here we notice the similarity between the interpretation of Newton and Einstein. For all practical purposes, the planetary oribts as determined by Einstein's criterion are the same as those in the Newtonian theory! There are very slight differences, which are most noticeable for the planet Mercury because it lies closest to the Sun and has the most eccentric of all planetary orbits. Einstein's theory predicts that Mercury's orbit slowly rotates in space at the rate of about 43 arc-seconds per century.

In our earlier discussion, we did come across such behavior on the part of planet Mercury. We had found that the perihelion of its orbit advances at the rate of about 43 arc-seconds per century— an effect that had remained unexplained by the Newtonian law. Now we see that Einstein's theory provides an explanation. In fact, its prediction agrees very well with the observed rate of advance.

This remarkable success of general relativity inspires confidence in the theory. It was, however, another astronomical observation that established the viability of the theory. This was the observation of the bending of light rays near the Sun.

Light, we know, travels in a straight line. If the definition of a straight line changes according to the system of geometry, the path of a light ray will be different in curved space than in flat space.

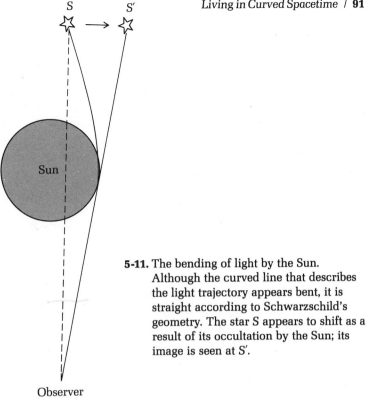

S   S'

Sun

5-11. The bending of light by the Sun.
Although the curved line that describes
the light trajectory appears bent, it is
straight according to Schwarzschild's
geometry. The star S appears to shift as a
result of its occultation by the Sun; its
image is seen at S'.

Observer

In Figure 5-11, we see two light paths passing close by the surface
of the Sun. The dashed straight line describes the path in Euclid's
geometry; the curved line is the path followed by light in Schwarzschild's geometry. To us, accustomed to Euclid's geometry, this
path appears bent. But it is in fact the path of a light ray *moving
straight* in the non-Euclidean geometry of Schwarzschild's solution. Which path does light actually follow?

Astronomers can answer this question by the following experiment. Suppose they observe a star that is occulted by the Sun, that
is, a star whose line of sight is crossed by the Sun. In the situation
where the Sun is away from the line of sight to the star, the difference between Schwarzschild's geometry and Euclid's geometry is
negligible, and the Euclidean straight line from the star to us represents the path of light from the star. In Figure 5-11, just before (or
after) the occultation by the Sun, the light ray from the star S grazes
the solar limb and should show the maximum bending, as indi-
cated by the curved line. The direction of the star should therefore

appear to change as it is occulted by the Sun. The expected change in the direction, according to Einstein's theory, is very small—about 1.75 arc-seconds.

What prediction can one make on the basis of Newton's law of gravitation? Newton's law in its original form did not envisage that light would be attracted by the Sun at all, and so the expected change of direction is zero. However, if one assumes that light is made of tiny particles (called *photons*) that are subject to Newton's law of gravitation in the same way as particles of matter are, then the change in the star's direction in the above experiment should be exactly half that predicted by Einstein's theory.

Here then is an experiment that can, in principle, tell us which interpretation—Newton's or Einstein's—is right or whether both are wrong. In practice there are several difficulties. First, we cannot see the star when the Sun is shining brightly in the sky. The only occasion when we can perform this experiment is when there is a total solar eclipse, not a very common phenomenon. Next, the expected change of direction is very small and requires very accurate measurement techniques. The situation is further complicated by the fact that, very near the Sun's surface, the hot gas can also bend the light rays just as a lens bends them.

As far as the first difficulty is concerned, astronomers did not have to wait long. In 1919, just four years after general relativity was proposed and only three years after the Schwarzschild solution, a total solar eclipse did take place. Realizing the importance of the experiment, the English astronomer Eddington took the initiative in making the measurements. Eddington and Cottingham went to Principe, an island in the Gulf of Guinea, while their colleagues Davidson and Crommelin went to Sobral in Brazil, places where the total eclipse could be seen. Thanks to a financial grant of £1000 obtained by the Astronomer Royal Sir Frank Dyson, these eclipse expeditions became possible.

The results of these observations favored Einstein's theory rather than Newton's. When Sir Frank Dyson announced these results to a crowded meeting of the Royal Society in London, they caused a considerable sensation. Later, A. N. Whitehead recaptured the scene in the following words: "The whole atmosphere of tense interest was exactly that of a Greek drama: we were the chorus commenting on the decree of destiny as disclosed in the development of a supreme incident. There was . . . in the background the picture of Newton to remind us that the greatest of scientific generalisations was now, after more than two centuries, to receive its first modification. . . ."

In retrospect we now see that, although the 1919 eclipse observations did go a long way toward establishing general relativity as a viable theory of gravitation, those observations were by no means conclusive. When an astronomer (or indeed any scientist) makes a measurement, his result is always subject to a number of imponderable or uncontrollable variations usually denoted as errors. Only when the errors are sufficiently small can one confidently assert that the reported measurement has confirmed or disproved a theoretical prediction. The range of errors of the Principe and Sobral eclipse observations was so large that the Newtonian prediction could not in fact be ruled out!

Present technology makes it possible to do this experiment much more accurately using microwaves rather than visible light. Instead of a star, the quasar 3C 279 (3C here stands for the *Third Cambridge Catalogue*) is observed in the microwave region of the spectrum as it is occulted by the Sun. In this case, the Sun's own radiation is so small that we do not have to wait for an eclipse. Also, the solar atmosphere does not distort the path of microwaves as much as it distorts the path of visible light. The measurements made in 1975–1976 by the astronomers at the National Radio Astronomy Observatory in Green Bank, West Virginia, were so accurate that it can now be confidently asserted that Einstein was right and Newton wrong.

Is time also curved?

In 1979, in Einstein's centenary year, astronomers made another remarkable discovery, shown in the photograph at the beginning of this chapter. Are the quasars 0957 + 561 A,B in the photograph two distinct objects, or are they images of the same quasar? The similarity of the spectra of the two quasars has led many astronomers to suggest the latter alternative. If this is indeed the case, we have an excellent example of bending of light rays by gravity. Here the bending could be achieved by an intervening galaxy that acts as a gravitational lens.

These examples illustrate how the effects of the curvature of spacetime can be measured by astronomers. The examples we have discussed seem to be related to measurements in space. Are there any tests that can tell us how time measurements are affected by non-Euclidean geometry? The answer to this question is yes, but we will wait until Chapter 7 to discuss this answer.

The rings and limb of Saturn, photographed by Voyager 1 on November 12, 1980, from a distance of 700,000 km. The rings are a striking example of the effects of Saturn's tidal force; if the particles that make up the ring system were located outside the planet's Roche limit, they probably would have formed a moon rather than a ring system. (NASA.)

6

Ocean Tides and Binary Stars

6

Ocean Tides and Binary Stars

When Newton and Einstein agree

Einstein's general theory of relativity and Newton's law of gravitation offer radically different interpretations of the phenomenon of gravity. Yet, in practical terms, the differences between their predictions seem to be very small. In Chapter 5 we saw two examples of observations in the solar system, the precession of the orbit of Mercury and the bending of light rays from a distant star by the Sun. In both cases the differences in the predictions of Newton and Einstein are very small and are measurable only with very patient and sophisticated astronomical observations. Is it just a coincidence that these two approaches give almost the same answer?

A mathematical analysis of Einstein's equations tells us that the agreement between the two approaches is not coincidental. It can be shown that, in *all* phenomena of *weak* gravitational effects, the two theories must almost agree. In our discussion of the escape speed in Chapter 3, we saw how to measure the relative strength of gravity. We use the criterion of the escape speed in the present context to understand the difference between "weak" and "strong" gravity. The rule is simple: compare the escape speed V with the speed of light c. If the ratio V/c is very small compared to 1, the gravitational effects are weak. If the ratio is very close to 1, say

Weak gravity   Strong gravity

6-1. When the escape speed V is small compared to the speed of light c —that is, when we are dealing with situations of weak gravity— Newton and Einstein agree. When V becomes comparable to c, the disagreement between the two theories becomes significant.

between 0.1 and 1, the gravitational effects are strong. Referring back to Table 3-2 of Chapter 3, we see that the gravitational effects are weak in all cases except on the surface of neutron stars.

This is the reason why, in spite of the conceptual and observational superiority of Einstein's theory, Newton's law is still usable. Indeed, because the mathematical formalism of general relativity is much more complicated than the Newtonian method, astronomers *prefer* to use the latter in cases of weak gravity. In this chapter, we shall describe a few phenomena of weak gravity. In our discussion, we shall refer to the Newtonian framework, except toward the end. In subsequent chapters, however, we shall be concerned mostly with *strong* gravity, and there we shall follow Einstein's framework. In discussing strong gravity, the difference between the approaches of Newton and Einstein becomes significant, and we should therefore use what we believe to be the more reliable of the two frameworks.

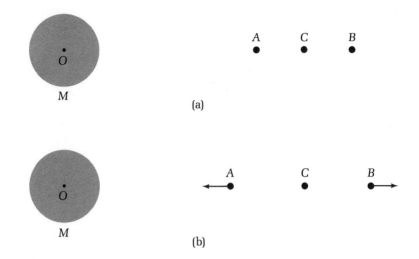

(a)

(b)

6-2. When a force of just the right magnitude is applied to hold *C* where it is, the same force is inadequate to hold *A* and more than adequate to hold *B* where they are. In either case, *A* and *B* move away from *C*.

The tidal force

Consider first the following situation illustrated in Figure 6-2. *M* is a gravitating spherical mass exercising its force of attraction on three equal point masses *A*, *B*, and *C*. As shown in Figure 6-2a, the masses *A*, *B*, and *C* are located in a straight line with *C* in the middle. To illustrate the point we want to make, we take the simple case of the line *ACB* passing through the center *O* of *M*.

From Newton's law, we know that *M* attracts all three of the masses *A*, *B*, *C* as if all its mass were concentrated at *O*. We also know that, by the inverse-square law, this force of attraction is the largest on *A* (which is closest to *O*) and the least on *B* (which is farthest from *O*). The tendency of this force is of course to make *A*, *B*, and *C* fall toward *O*.

Now suppose that we wish to check this tendency by applying a force in the opposite direction. Let us apply an equal and opposite force on *C* that just holds it at rest. If we apply the *same* force on *A*, it will not be sufficient to counter the force of gravity from *M*. As a result, *A* *will still move toward* *O*. However, if we apply the same force on *B*, it will exceed the force of gravity on *B*, as a result

Moon

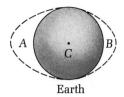

Earth

6-3. The dashed lines indicate the Earth's tendency to bulge at ends *A*
and *B* along the line joining the centers of the Earth and the Moon.
Compare this situation with that described in Figure 6-2. In both cases
the tidal effect is larger at *A* than at *B*.

of which *B will move away from O.* In other words, the masses *A*
and *B* both move away from *C.* The length *AB* tends to increase,
as shown in Figure 6-2b.

If the particles *A*, *B*, and *C* were connected and were part of the
same body, the effect of our example would be to stretch the length
AB. This stretching force is known as the *tidal force* exerted by *M*
on the body. Besides exerting a force of attraction, *M* also tends to
distort the shape of the body through this tidal force.

Tides in the oceans

The name *tidal force* has come from its application to the tides
on the Earth. In the example just discussed, let *M* stand for the
Moon and let the system of three particles be replaced by the Earth.
We then encounter the situation shown in Figure 6-3.

We see in Figure 6-3 that the dashed line shows the tendency of
the Earth to become elongated along the line toward (and away
from) the center of the Moon. The rigidity of the Earth's solid crust
prevents a great bulging in response to the tidal force. However,
the effects of this force do show in the motions of the ocean tides.

The effect is naturally magnified if the Sun also happens to lie
approximately in the Earth–Moon line. This happens on the occasions of the Full Moon and the New Moon. On these occasions, the
tides in the ocean are the most spectacular.

Where does the energy of motion of the water come from in an
ocean tide? Obviously, the source of this energy lies in the tidal

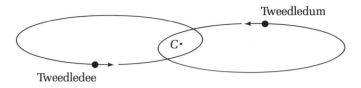

6-4. The highly eccentric elliptical orbits of Tweedledum and Tweedledee. They move in such a way that their center of mass, which is the point C midway between them, remains fixed in space.

force of gravity exerted by the Moon on the Earth. In these days of energy shortages, we cannot afford to ignore natural sources of energy, and the ocean tides represent one such natural source.

The experiment of Tweedledum and Tweedledee

In fact, the tidal force is a mechanism whereby energy can be transferred from one astronomical object to another. An interesting thought experiment illustrating this process was once described by two astrophysicists, Hermann Bondi and William McCrea. The experiment concerns two remarkable creatures called Tweedledum and Tweedledee. They are made of a pliable material that allows them to change shape. Originally, they had identical spherical shapes. They were set moving around each other in highly eccentric orbits under their mutual gravitational force. They were given an order that they must always move along these orbits. Let us assume further that both Tweedledum and Tweedledee always have symmetrical shapes about axes perpendicular to the plane of their orbits.

Figure 6-4 illustrates the orbits of Tweedledum and Tweedledee. The orbits are identical, and the two move along their respective orbits in such a way that the point midway between their centers is fixed. This is characteristic of the way any two objects move in each other's gravitational attraction. Even in the case of the Earth and the Sun, although we usually talk of the motion of the Earth only, the Sun should in principle also move under the Earth's gravitational force. However, in the Earth–Sun system, because the

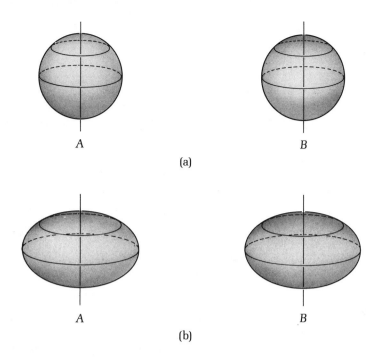

6-5. In (a) we have two spherical objects A and B. Because of their gravitational tidal force on each other, both tend to bulge in the middle. If they are constrained to remain symmetric about an axis as in (b), they become oblate spheroids.

Sun's mass greatly exceeds that of the Earth, the Sun's motion is negligible. The center of mass of the Earth and the Sun in fact lies within the Sun.

But to return to Tweedledum and Tweedledee: they would have no difficulty obeying the command to move along their assigned orbits if they were rigid creatures. Since they are pliable, their shapes get distorted by each other's tidal forces. Let us investigate this effect further.

In Figure 6-5a, we have two spheres A and B. We saw earlier that, if B is subject to A's tidal force, it will bulge in the direction of A. If B is also constrained to be symmetrical about an axis perpendicular to this direction (as Tweedledum and Tweedledee are), it will bulge all along its equator, as shown in Figure 6-5b. The same happens to A under the tidal force of B.

When a sphere bulges along its equator as in Figure 6-5b, it becomes an *oblate spheroid*. Had B instead been elongated at the poles, it would have become a *prolate spheroid*, as shown in

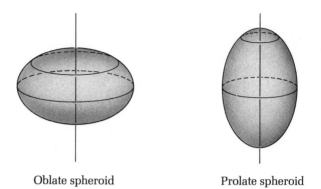

Oblate spheroid   Prolate spheroid

6-6. An oblate and a prolate spheroid.

Figure 6-6. An oblate spheroid is bun-shaped, and a prolate spheroid is more egg-shaped.

In Figure 6-5b, when the two spheres become spheroids their mutual gravitational attraction *increases*. This is because A attracts B more powerfully as an oblate spheroid than as a sphere. This is the reason why the job of Tweedledum and Tweedledee is made more difficult—as they go around each other, their shapes tend to become oblate and the force between them increases, an effect which in turn tends to distort their orbits. To overcome this difficulty, one course is open to them. If one of them becomes oblate, the other must become prolate because the prolate spheroid exerts less force on its companion than a sphere.

Now Tweedledum is clever and unscrupulous whereas Tweedledee is stupid and simple. Tweedledum uses his intelligence to capitalize on the situation in the following way. He makes Tweedledee sign an unfair agreement saying that whenever Tweedledum becomes oblate, Tweedledee must become prolate, and vice versa. The exact shape that Tweedledee must acquire will be calculated and communicated to him by Tweedledum. The basis of the calculation is the criterion that their orbits must not change. Not knowing how to make this calculation himself, Tweedledee simply signs along the dotted line. This is a mistake for which he will have to pay dearly, as we will now see.

Moving in highly eccentric orbits, Tweedledum and Tweedledee alternately come close together and move farther apart. In Figure 6-7a, they are shown near each other. At this stage, there are huge

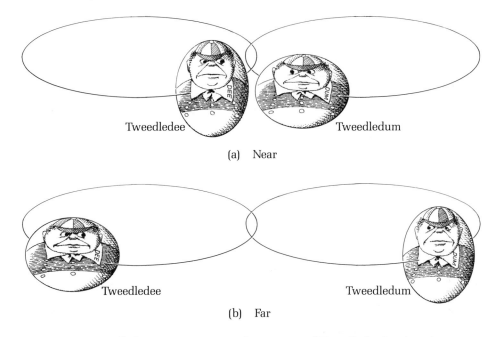

Tweedledee   Tweedledum

(a)   Near

Tweedledee   Tweedledum

(b)   Far

6-7. Tweedledum gains energy at the expense of Tweedledee by changing
shape in accordance with tidal forces.

tidal forces between them, and here Tweedledum becomes oblate.
To achieve this, he of course does not have to do any work; as
shown in Figure 6-5, the tidal force does the work of changing his
shape. And this work done by an outside force increases Tweedledum's energy reservoir. What about Tweedledee? He must go into
an appropriate prolate shape as per the agreement he has signed.
To do this, he has to work against the tidal force, and as a consequence his energy reservoir is depleted.

Meanwhile, the two have started moving apart. When their separation is close to the maximum, as in Figure 6-7b, Tweedledum
changes himself to a prolate shape, thereby making Tweedledee go
oblate. However, when the two are far apart, the tidal force between them is weak. Hence, although Tweedledum does some
work to become prolate, he has to spend very little energy doing so.
Correspondingly, Tweedledee gains some energy because this time
the tidal forces *help* him change his shape. But his energy gain
is small.

We now see how unfair this agreement is. During each orbit,
Tweedledum alternately gains a lot of energy and spends a little,

while Tweedledee loses a lot of energy and gains a little. What is happening is that Tweedledum is cleverly using the tidal force to extort energy from poor Tweedledee, although the latter cannot, on the face of it, see where the unfairness of the agreement lies.

Tidal disruption

The crucial property of the tidal force that leads to the remarkable effects of the Tweedledum–Tweedledee experiment is that its strength diminishes with distance very rapidly. In fact, its strength diminishes in inverse proportion to the *cube* of the distance. Thus, if the farthest distance of separation between Tweedledum and Tweedledee is ten times the distance of their closest separation, the tidal force in the former case is one-thousandth of what it would be in the latter case. This is why Tweedledee had to work so much harder than Tweedledum.

We therefore expect the tidal force to be very powerful in circumstances where two astronomical objects are close to each other. How close can a planet be to its parent star? How close can a satellite come to the planet around which it revolves? The answers to these questions invariably involve calculations of the tidal force. If a satellite comes too close to a planet, the tidal force may become so enormous that it destroys the satellite. The same consideration applies to a planet near its parent star. The limit beyond which the tidal forces become highly disruptive is known as the *Roche limit*.

Another circumstance in which the tidal force becomes significant is for a binary star system. Like Tweedledum and Tweedledee, two stars in a binary system go around each other in elliptical orbits. In Figure 6-8a is shown a typical binary system. The stars *A* and *B* (which need not be and usually are not of identical mass) orbit around their common center of mass. The dashed line shows the so-called *Roche lobe*. If either star becomes large enough to cross its Roche lobe, the tidal force of its companion will begin to make its disruptive presence felt.

We saw briefly in Chapter 4 that, as a star evolves, it passes through the red-giant phase and becomes very large. If one of the stars in our binary system becomes a red giant, its outer surface begins to spill over its Roche lobe, and the situation shown in

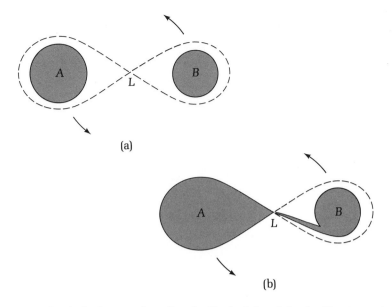

(a)

(b)

6-8. The dashed curve describes the Roche lobe of the double star system. The point L is called a Lagrangian point. In (a), stars A and B are well inside their Roche lobe. In (b), star A has expanded and filled its Roche lobe. The tidal forces of B now pull out material from A, which escapes through L and falls into B. The arrows indicate the direction in which the binary system is rotating.

Figure 6-8b will occur. The companion star B now starts exerting a tidal force on A in such a way as to pull material from the surface of A that faces it. This surface therefore becomes distorted, and the material "falls" toward B. We will return to this interesting situation in Chapter 8, for it has potentially dramatic consequences.

Gravitational radiation

The effects we have been discussing so far could be analyzed within the framework of general relativity in much the same way that they are analyzed within the Newtonian framework. We have used the Newtonian framework to describe these effects as it is intuitively the simpler of the two to grasp. Now, however, we turn to another effect of weak gravity that can only be described within the relativistic framework. This is the phenomenon of gravitational radiation.

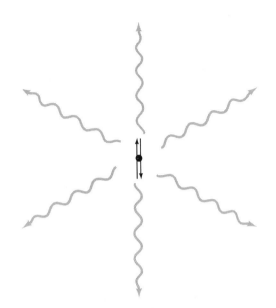

6-9. A schematic representation of an electric charge. Such a motion generates disturbances of electricity and magnetism around the charge. These disturbances move as waves in directions radially away from the moving charge.

Recall from Chapter 5 the criticism of Newton's law of gravitation, that gravitational effects supposedly propagate instantaneously, that is, with *infinite speed*. According to Einstein's theory of special relativity, no physical effects can be communicated across space faster than the speed of light. Does Einstein's theory of gravitation as stated in the general theory of relativity meet this requirement?

The answer to this question is yes, but the details of how the gravitational effects propagate from *A* to *B* are immensely complicated when we are dealing with the situations where these effects are strong. However, for weak gravitational effects, the situation is considerably simpler and is analogous with the more familiar case of electromagnetic radiation. Like electromagnetic radiation, one can talk of *gravitational radiation*.

The simplest system generating electromagnetic radiation is one where an electric charge oscillates, or moves to and fro much like the bob of a simple pendulum. As shown in Figure 6-9, the radiation emanating from such a motion of the electric charge consists

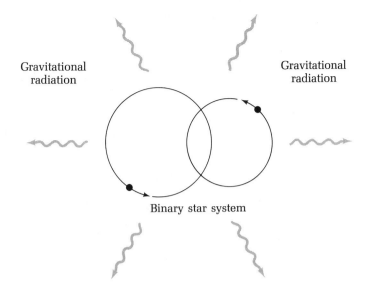

6-10. Schematic representation of gravitational radiation by a binary star system.

of electrical and magnetic disturbances that propagate away from the charge *with the speed of light*. The disturbances generated on the initially calm surface of a pond of water when a pebble is dropped into it are roughly analogous to electromagnetic radiation.

The electrical and magnetic disturbances carry energy. Where does this energy come from? Its sources can be traced to the motion of the charge. As it radiates electromagnetic waves, the electric charge tends to lose its kinetic energy and hence experiences a damping of its motion. Just as the bob of the pendulum eventually slows to a state of rest because of the damping effect of air resistance, our oscillating charge also slows down because of the damping produced by radiation.

The gravitational analogue of this situation is described in Figure 6-10. This is the now-familiar binary star system. As the stars go around each other, they generate disturbances in the geometry of spacetime. These disturbances propagate out to large distances *with the speed of light*. And as in the case of the electric charge, these disturbances carry energy, which causes a damping of the binary star motion. Because of this damping, the stars come closer and closer (their orbits shrink), and the angular speed with which they go around each other increases.

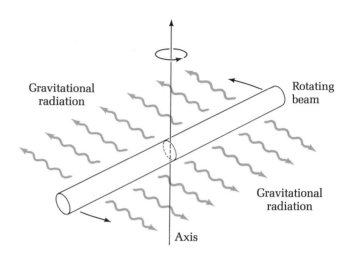

6-11. A gravitational radiator. The cylindrical beam rotating about an axis perpendicular to its length will radiate gravity waves. The effect is, however, very, very small.

Has gravitational radiation been detected in a laboratory experiment? Here we begin to depart from our analogy with electromagnetic radiation. Unlike the electromagnetic case, gravitational radiation is so feeble that production of gravity waves lies beyond the capabilities of the terrestrial laboratory. To get some idea of the problem, imagine a cylindrical beam of radius 1 meter and length 20 meters, density 7.8 times that of water and a mass of about 490 tons (see Figure 6-11). To produce gravitational radiation, we rotate this beam about its middle with an angular velocity of just under 4.5 revolutions per second. To generate gravitational radiation of 1 watt of power, we need about 1 million million million million million such cylindrical sources!

There is, however, some indirect evidence for the existence of gravitational radiation from astronomical sources. In early 1979, J. H. Taylor, L. A. Fowler, and P. M. McCulloch, observing from the 1000-foot dish at the radio astronomy observatory in Arecibo, Puerto Rico, reported that their monitoring of the motion of the binary pulsar PSR 1913 + 16 (consisting of the pulsar and another

compact star) shows the effects expected from graviational radiation.* In particular, their orbital period around each other *decreases*
at just the rate expected on the basis of gravitational radiation. This
interpretation is still subject to some uncertainty until we can
definitely rule out any other plausible explanations of these observations. Therefore, until the laboratory detecting devices improve
considerably, we have to be content with such indirect evidence
for gravitational radiation. Considerable improvement is needed in
the existing laboratory detectors of gravitational radiation before a
direct measurement of radiation from such astronomical systems
becomes feasible.

The binary pulsar PSR 1913 + 16 is of interest to relativists for
another reason. Just as in the case of the progression of the perihelion of the planet Mercury (see Chapter 5), we see here a similar
effect. The direction along which the two orbiting stars come
closest to each other is found to change with time. (The Newtonian
law of gravitation predicts this direction to be stationary.) The observed rate of change of this direction is considerably faster than in
the case of Mercury; it is about 4.2° per year. If it becomes possible
to estimate the masses of the two stars accurately, we can then
compare this observed precession rate with a theoretically predicted rate.

In any case, PSR 1913 + 16 is another example of how accurate
astronomical measurements made possible by our present technology can detect even the weak effects of gravity.

---

*The letters PSR indicate that the source is a pulsar. The numbers following
these letters fix the direction of the pulsar in the sky.

The Crab nebula is the remnant of a supernova explosion that occurred in the year AD 1054. This X-ray photograph shows the pulsing neutron star that formed from contraction of the central core of the star. In the left photograph, the pulsar is "on" (beaming light in the direction of Earth); in the right photograph, the pulsar is "off." The neutron star pulses ~30 times per second. (Courtesy of the Einstein Observatory/Harvard-Smithsonian Center for Astrophysics.)

7

The Strange World
of Black Holes

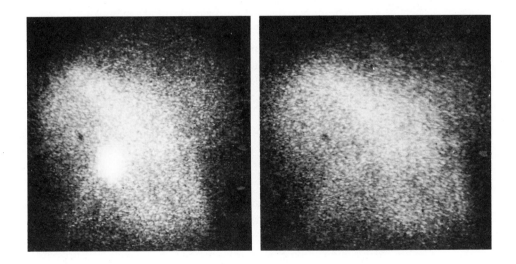

7

The Strange World
of Black Holes

The black hole in history and astronomy

The oldest mention of a black hole is found not in books of physics or astronomy but in books of history. In the summer of the year 1757, Nawab Siraj-Uddaula, the ruler of Bengal in eastern India, marched on Calcutta to settle a feud with the British East India Company. The small garrison stationed in Fort William at Calcutta was hardly a match for Nawab's army of 50,000. In the four-day battle that ensued, the East India Company lost many lives, and a good many, including the company's governor, simply deserted. The survivors had to face the macabre incident now known as the Black Hole of Calcutta.

The infuriated Nawab, whose army had lost thousands of lives in the battle, ordered the survivors to be imprisoned in the Black Hole, a prison cell in Fort William. In a room 18 feet by 14 feet, normally used for housing three or four drunken soldiers, the 146 unfortunate survivors were imprisoned. The room had only two small windows (see Figure 7-1). During the 10 hours of imprisonment, from 8 PM on June 20 to 6 AM on June 21 in the hottest part of the year, 123 prisoners died. Only 22 men and one woman lived to tell the tale.

Apart from its macabre aspect, the Black Hole of Calcutta does bear some similarity to its astronomical counterpart, involving as it did a large concentration of matter in a small space from which there was no escape. While discussing the notion of escape speed in Chapter 3, we made a passing reference to the concept of a black hole. We now define it as an object whose gravitational attraction is so strong that not even light can escape from its surface.

7-1. The Black Hole of Calcutta. (From Noel Barber, *The Black Hole of Calcutta*. London: Collins.)

The French mathematician Laplace conceived of the notion of a black hole in the year 1799. Although Laplace did not use the term *black hole* to describe these objects, it is clear from his discussion that his concept implied the property of the escape speed exceeding the speed of light. For a spherical object of mass M and radius R, this property implies that R cannot exceed the value

$$R_s = \frac{2GM}{c^2}$$

where c is the speed of light and R_s is the critical value for the radius of a Laplacian black hole. (The meaning of the subscript s will become clear when we discuss black holes within the framework of general relativity.)

To get an idea of how small R_s is, let us look at a couple of examples. Suppose we shrink the Earth until it satisfies the above criterion for a black hole. In this case, $R_s \cong \frac{1}{3}$ inch! That is, the radius of the Earth must be less than a third of an inch if it is to become a black hole. If we consider a somewhat more massive object, the Sun, the corresponding value of R_s is about 1.5 miles. The actual radius of the Sun is nearly 300,000 times R_s.

It appears, however, that Laplace was not the first to talk about black holes in this way. An English physicist named John Michell published a paper in 1784 in the *Philosophical Transactions of the Royal Society of London* (Vol. 84, p. 35), with the title "On the Means of Discovering the Distance, Magnitude, etc. of the Fixed Stars, in Consequence of the Diminution of the Velocity of the Light, in case of such a Diminution should be found to take place in any one of them, and such other Data should be procured from Observations, as would be further Necessary for that Purpose." In this paper Michell discussed stars in general and speculated about the possibility of astronomical objects whose mass and radius satisfy the criteria for a black hole.

These speculations, and the work of Laplace, were based on the Newtonian theory of gravitation. We have already seen that, when discussing effects of strong gravity, the Newtonian framework is suspect and that modern physicists prefer to use Einstein's general relativity theory. Let us therefore discuss the phenomenon of black holes within Einstein's framework. We will find that black holes are even more dramatic within this framework than within the Newtonian framework.

How are black holes formed?

It is already clear from the two examples of the Earth and the Sun that even within the Newtonian framework black holes are highly unusual objects. Even astronomers accustomed to dealing with extreme states of matter find black holes esoteric.

As we saw in Table 3-2, the stronger the gravitational attraction of an object, the higher is the velocity required to escape from it. For stars and planets, the escape speed is a small fraction of the speed of light. Even for neutron stars, the escape speed does not exceed about two-thirds the speed of light.

The neutron stars, however, suggest a clue to the formation of black holes. In a neutron star, the density of matter reaches a value as high as a *million billion times* the density of water. What force is responsible for holding a neutron star in equilibrium? We have seen in Chapter 4 how powerful the force of gravity is inside a star.

7-2. The remnant of a supernova explosion may be a neutron star.

The Sun, for example, would not be able to maintain its present size against the contracting tendency of its own gravitational force if it were not for the thermal pressures within it. The thermonuclear reactions that take place in the deep interior of the Sun not only generate enough energy to keep the Sun shining but also provide a stabilizing outward pressure that keeps the Sun from collapsing. In a neutron star, no thermonuclear reactions take place. In fact, the neutron star is formed toward the *end* of a star's normal lifetime, when it explodes as a supernova.

In Chapter 4 we carried the story of stellar evolution to this supernova stage. We did not ask what is left behind when the star explodes as a supernova since the question is of greater relevance in the present context than in our discussion in Chapter 4 of the star's role as a fusion reactor.

Our present understanding of the supernova explosion process is by no means perfect, but theoretical work in this area of astrophysics suggests that the central hot core of the star is left as a remnant (see Figure 7-2). This core contracts and becomes a neutron star *if its internal pressures can support the crushing force of the star's own gravity.*

What are the internal pressures in a neutron star? At the high density of up to a million billion times the density of water, the matter in the star exists predominantly in the form of neutrons. And these neutrons are closely packed to 10^{40} neutrons per cubic inch.

The laws of quantum physics describe the behavior of matter at the very small scale of atoms and nuclei. These laws describe certain restrictions on the way a set of identical particles can be packed in a given volume. When these restrictions (first discovered by the atomic physicist Wolfgang Pauli) are calculated for neutrons packed inside a neutron star, we find that this matter is endowed with a new kind of pressure. This pressure has a tendency to *resist* closer packing of neutrons. This outward pressure *attempts* to maintain the neutron star's equilibrium against its self-gravity.

Does it seem that we have digressed somewhat from our question of how black holes are formed? We seem to have arrived instead at a description of how a neutron star is formed and maintained in equilibrium. However, we have not digressed from the topic of black holes! We are in fact very close to the scenario that leads to the formation of a black hole.

In our discussion of the internal pressure of a neutron star, we said that this pressure *attempts* to maintain equilibrium. The attempt does not always succeed! There are limitations on how much weight this type of pressure can support. Calculations show, for example, that if the mass of the neutron star exceeds three times the mass of the Sun (3 $M_\odot$), this pressure cannot maintain equilibrium. The mass limit of 3 $M_\odot$ may in fact be an overestimate —it could be as low as 2 $M_\odot$.

What happens if the remnant of a supernova explosion has a mass exceeding this limit? The remnant obviously cannot survive as a neutron star. Its pressures will be inadequate to withstand the crushing force of its own gravity. The star continues to contract and becomes a black hole.

Gravitational collapse

Let us examine the process of the contraction of a star to a black hole in some detail. In Figure 7-3, we see a contracting star in two stages. In Stage I, the star has just begun to contract because

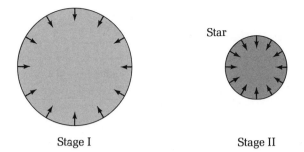

Stage I · Stage II

7-3. In Stage I, gravity is beginning to assert its supremacy over the internal pressures in the star. In Stage II, gravity has far surpassed the internal pressures and is dominating the dynamical behavior of the star. The star now undergoes a gravitational collapse.

its internal pressures have proved inadequate to balance the inward force of gravity. The rate of contraction here is slow. In Stage II, at a later time, the star is considerably smaller. According to the Newtonian framework, the force of gravity within the star will grow considerably stronger as it shrinks from Stage I to Stage II because all its matter has come closer together. Although the star's outward pressure might have grown during this contraction, the outward pressure increases more slowly than the star's inward force of self-gravity. In other words, the imbalance between the two opposing forces in Stage I has increased in Stage II. The star therefore contracts even faster than before, *and this tendency to contract will increase as the star contracts further.*

This situation is often described by the term *gravitational collapse.* From the initial gentle rate of contraction, the star progresses to a catastrophic implosion when its self-gravity becomes so strong that nothing can prevent the collapse of the star.

Of course, we used the Newtonian framework of gravity to explain what is meant by gravitational collapse. Since we are dealing with very strong gravitational effects, we should in fact have used general relativity. Had we done so, in qualitative terms the behavior of the star would have turned out to be no different. Relativity, however, introduces certain novel features into this situation that we can no longer ignore.

Gravitational redshift

Let us recall that our discussion of the non-Euclidean geometry of spacetime in Chapter 5 was limited to measurements in space only. This is an opportune time to describe the effects of non-Euclidean geometry on time measurements.

Figure 7-4a shows a spacetime diagram in the imaginary situation where *no gravity* exists—that is, where spacetime is flat. The lines *a* and *b* are the world lines of two observers at rest, A and B. Suppose B sends light signals to A every second as measured by his own clock. The dashed lines describe the tracks of light rays leaving B and reaching A. At what intervals does A receive these signals? These light tracks and the world lines *a* and *b* form a succession of parallelograms. In any parallelogram, the opposite sides are equal, and so as measured by A's clock the signals from B will arrive at one-second intervals.

This conclusion is based on Euclid's geometry and does not apply to the situation illustrated in Figure 7-4b, which shows curved spacetime around a massive spherical object. The geometry of curved spacetime was first determined by Karl Schwarzschild (see Chapter 5). The shaded region indicates the presence of the massive object. The signal emitter B is now on the boundary of this object, while A is far away. As before, A and B are at rest relative to the massive object, and the intervals along their respective world lines *a* and *b* denote the times measured by their respective clocks. The dashed lines denote the light tracks from B to A.

On the plane of the paper, these tracks do not appear straight. However, the rules of geometry are no longer those of Euclid. By the rules of Schwarzschild's geometry, these dashed lines are *straight* light tracks. However, their intercepts on A's world line are longer than those on B's world line. This difference of length means that, although B sends signals at one-second intervals by his clock, A has to wait *longer* than one second, say $(1+z)$ seconds, between successive signals from B. This fractional increase of z is known as *redshift*.

Suppose that, instead of sending signals to A every second, B sends light waves of a specific frequency v all the time. This means

Euclidean geometry

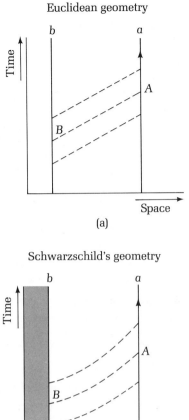

(a)

Schwarzschild's geometry

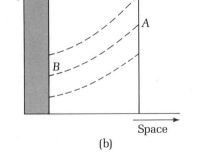

(b)

7-4. Two examples of spacetime geometry. (a) Euclid's geometry, which applies in the absence of gravity, and (b) Schwarzschild's geometry, which applies near a massive spherical object, whose boundary is shown by the shaded area. In both cases, time intervals are measured by the observers A and B, whose world lines are shown by a and b, respectively. The dashed lines are tracks of light signals emitted by B at regular intervals.

that in every one-second interval, ν wave crests will leave B toward A. These same wave crests would be received by A in the interval $(1+z)$ seconds. In other words, the frequency of waves received by A is reduced to $\nu/(1+z)$.

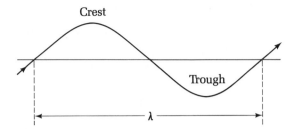

7-5. The profile of a typical wave. From the level position, the wave rises to the maximum height, the crest, then passes through the level position down to the minimum height, the trough, and then rises back to the level position. The wavelength λ measures the distance between the initial and the final level positions. It also measures the distance between successive crests or troughs of the wave. If in one second ν wave crests pass a given point, the wave will have advanced through a total distance $\nu \times \lambda$. This is the speed of the wave.

There is a simple relation (illustrated in Figure 7-5) between the frequency (ν, nu) and wavelength (λ, lambda) of a light wave:

$$\text{frequency} \times \text{wavelength} = \text{speed of light}$$
$$\nu \times \lambda = c$$

In Figure 7-4b, the frequency of the light wave is *reduced* by the factor $(1+z)$ in going from B to A. The wavelength is therefore *increased* by the factor $(1+z)$. That is, if λ_B and λ_A are the wavelengths of the light wave as measured by B and A, respectively, we have

$$\lambda_A = (1+z) \times \lambda_B$$

If B were emitting a whole spectrum of visible light (as a star does), A would receive that spectrum with all its wavelengths systematically increased by this factor $(1+z)$. Of all the colors that make up the visible spectrum, red light has the largest wavelength. Therefore, relative to B's spectrum, A's spectrum will have shifted toward the red end. Hence, z is named the *redshift*. Since this effect has been caused by the gravity of the massive object, it is known as the *gravitational redshift*.

This effect is very small for the Sun. In general, for weak gravity, we have a simple formula for z:

$$z = \frac{GM}{c^2 R}$$

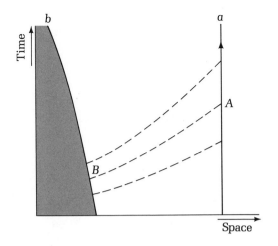

7-6. Signal propagation between *A* and *B* in the case where *B* is
located on the surface of a collapsing object (*shaded area*).
Although *B* sends signals at regular intervals, they reach *A*
at progressively longer intervals as measured by *A*'s clock.

This is the redshift of light leaving the surface of an object of mass
M and radius *R*. For the Sun, z is as small as two parts in a million.

For some other stars, this effect is somewhat larger. The best
case, from a practical point of view, is that of a *white dwarf* star.
Here z lies in the range from 10^{-5} to 10^{-3}. For example, for the white
dwarf Sirius B, the gravitational redshift is as high as 300 parts in a
million. Small though it is compared to 1, the gravitational redshift
is an indication of the effect of non-Euclidean geometry on time
measurements.

The event horizon

We now come to the more dramatic aspects of the gravitational
redshift where the gravitational effects are strong. Let us consider
a star undergoing gravitational collapse. We station an observer *B*
on the star's surface and instruct him to send out signals every
second. We station another observer *A* far away from the star and
along the radially outward direction from *B*. The situation, illustrated in Figure 7-6, is similar to that of the gravitational redshift
from the surface of a white dwarf star. The only difference between the white dwarf and the present collapsing star is that the

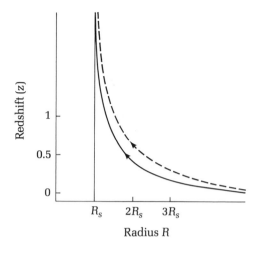

7-7. Schematic representation of how the redshift from the surface of a collapsing star increases as the star contracts. The solid curve describes the redshift increase as the star contracts *slowly*. The broken curve describes the redshift increase in the case where the star undergoes a rapid gravitational collapse. R_s is the Schwarzschild radius of the star. The redshift becomes infinite as R, the radius of the star, approaches R_s, for both the broken and the solid curves.

surface on which B is situated is collapsing. As the star collapses, B successively encounters an increasing force of gravity.

From our formula for the gravitational redshift, we find that as the star contracts its radius R *decreases* and the value of the redshift z *increases*. There is, however, one important difference from our previous example of the white dwarf. This formula is valid in situations of *weak* gravity. Our collapsing star with its decreasing radius will soon encounter *strong* effects of gravity. In such cases, this formula is changed to the following rather complicated form:

$$z = \frac{1}{\sqrt{1 - \dfrac{2GM}{c^2R}}} - 1$$

The solid curve in Figure 7-7 illustrates how the redshift changes with the radius R according to this formula. The redshift actually observed by A will be greater than that given by the formula; it is shown by the broken curve of the graph. This occurs because B is not stationary with respect to A but *is moving away from A.*

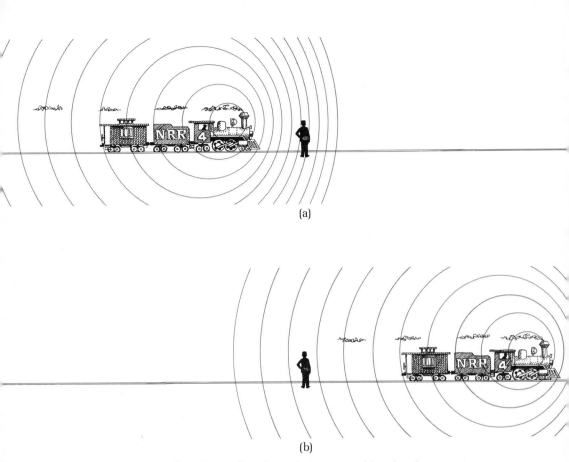

(a)

(b)

7-8. The Doppler effect. When the train is approaching the observer, the sound waves emitted by the whistle of the engine are closer together. This is shown in the figure by the narrow spacing of successive wavefronts as the waves move from the engine to the observer. The opposite effect takes place as the engine recedes from the observer.

Whenever a source of light moves away from an observer, the light from the source as seen by the observer is redshifted. This phenomenon is known as the *Doppler effect*. The Doppler effect is observed for sources of sound also. In Figure 7-8, we see an observer standing at a railroad crossing while a train passes by. The whistle of the engine appears to this observer to be *shrill* while the train approaches him and *flat* as the train recedes from him. The shrillness and flatness in this example indicate increase and decrease in the frequency of sound waves. The corresponding effects for light waves are known as *blueshift* and *redshift*. In our example of the collapsing star, *B*'s signals are redshifted because of both the Doppler effect and gravity.

Both the solid and broken lines of Figure 7-7 curve upward as *R* approaches the quantity

$$R_s = \frac{2GM}{c^2}$$

indicating that the redshift becomes infinitely large at $R = R_s$. This significance of R_s first became clear from the work of Schwarzschild, and for this reason R_s is often called the *Schwarzschild radius*. (The subscript *s* denotes this.) As *R* approaches R_s, *z* becomes infinite. To understand the meaning of this statement, recall that the factor $1 + z$ denotes the ratio of the rates at which the clocks of *A* and *B* run. The time interval between successive signals appears as one second to *B* and as $(1+z)$ seconds to *A*. As $1+z$ increases, *A* has to wait longer to get the next signal from *B*. When $1+z$ becomes infinite, *A* will have to wait *forever* to receive the next signal from *B!* In other words, as *B* crosses the sphere of Schwarzschild radius $R = R_s$, no signals from *B* will ever reach *A*. This is illustrated in Figure 7-9a.

Interpreted this way, the surface of the sphere of radius R_s constitutes an *event horizon*. No events taking place within the event horizon will ever be observed by an outside observer like *A*. The analogy here is with the horizon on the Earth. As shown in Figure 7-9b, the curvature of the surface of the Earth limits the range of its visibility from any given height. A ship observed on the high seas from a lighthouse disappears from view when its distance exceeds this range. Just as the ship is not visible beyond the Earth's horizon, so is *B* not visible once it crosses the event horizon.

The situation we have just discussed has an amusing analogy in everyday life. Imagine *A* to be an applicant for some financial grant from a highly bureaucratic government agency. *B* is the bureaucrat handling *A*'s correspondence. The time scales of *A* and *B* are, however, different. Whereas *A* expects a reply to his query within a day, *B* may take several days to oblige him with an answer. To *B*, whose normal existence is tuned to the red tape in his office, this delay would appear quite normal. We may call this phenomenon *bureaucratic redshift*. In a survey conducted by the admini-

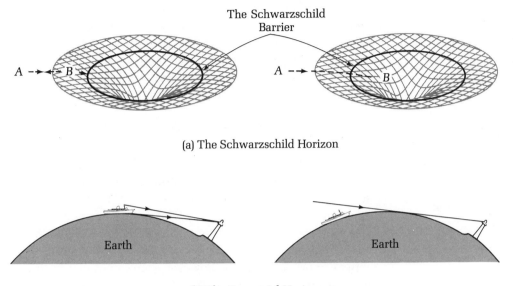

(a) The Schwarzschild Horizon

(b) The Terrestrial Horizon

7-9. Two horizons: (a) In the figure at left, A and B are able to communicate with each other along the two-way track shown with arrows. B is, however, moving away from A into a region of increasingly curved spacetime. In the figure at right, B has crossed the Schwarzschild barrier (shown by a thick closed curve). Once this happens, no signal from B reaches A. B can, however, receive signals from A. This barrier is the Schwarzschild horizon. In (b) the terrestrial horizon limits visibility due to the curvature of the Earth's surface.

stration of a state in India, it was found that a typical letter waits in the files of the bureaucrats for an average period of 288 days before receiving a reply! If we set our time scale ratio of $(1 + z)$ equal to 288, the bureaucratic redshift in this case equals 287.

But to return to our collapsing star: its gravitational redshift exceeds even this high value as it collapses further. When the surface of the object crosses the event horizon (that is, when its redshift becomes infinite), it becomes a black hole.

A black hole has no hair!

Can the outside observer A say when B has disappeared from view? Strictly speaking, A may wait *forever* but still not actually be able to say "Now I know that the star has become a black hole." The signal emitted by B just at the moment when he crosses the event horizon is destined never to reach A. Information about the history of B *prior to* this crucial instant is, however, accessible to A if he waits long enough.

Astronomers and astrophysicists often say that there is a black hole in such and such astronomical system. How can these statements be justified in view of what we have just said? Surely, any person on the Earth viewing an astronomical event is like the observer A. Never during his lifetime (however long that may be) can he claim that a black hole has been formed. In this exact sense, the claims made in the popular or technical literature about the existence of black holes are false.

In an approximate sense, however, such claims may be justified. Even before the star has actually crossed the event horizon, it may for all practical purposes become invisible to the external observer A because the gravitational redshift also acts in a way that drastically lowers the luminosity of the star.

We have seen how the frequency of light emitted from the surface of the star is reduced by the factor $(1+z)$ by the time it reaches A. The energy contained in a beam of light of a given frequency can be measured by first counting the number of *photons* in the beam (see Figure 7-10). A photon is an elementary particle that carries light in the form of an *energy packet*. Each packet has a certain energy, calculated by multiplying its frequency by a universal constant known as Planck's constant (usually denoted by h). So, as the beam of light travels from B to A, *its energy is reduced by the same factor by which its frequency is reduced*. This reduction factor $(1+z)$ becomes very large as the star's surface approaches the event horizon.

Because of this drastic reduction in its energy output, the star appears *almost black* even before it has reached the event horizon. Since all astronomical detecting instruments have thresholds of energy flux below which they cannot detect the sources of radiation, as soon as the collapsing star's energy flux drops below

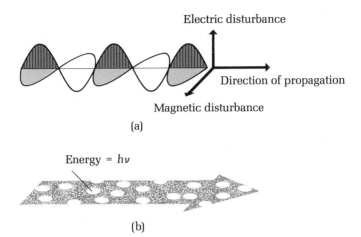

7-10. Light has a dual nature. In (a), light is shown as a wave with electric and magnetic disturbances propagating in wave patterns in mutually perpendicular directions. The positive disturbances are shaded; their maximum values correspond to wave crests. The number of crests per second equals the frequency (ν) of light. In (b), light is shown as made of packets of energy called photons. A photon of light of frequency ν has energy $h \times \nu$, where h is known as Planck's constant.

those thresholds, it will cease to be seen, even though it has not yet crossed the event horizon. In this approximate sense, the star may be said to have become a black hole. It is in this sense that we will henceforth use the statement that the star has become a black hole.

Even if a star has become a black hole in this way—that is, even if its radiation is too faint to be seen—its existence an be inferred by other indirect means. In particular, its gravitational effects will continue to be present. If the star had planets orbiting around it, these planets would continue to go around in their old orbits. In Chapter 3, we saw how we can estimate the mass of the Sun from information about the size and the period of the Earth's orbit. By similar techniques, we can in principle estimate the mass of a star even it it has become a black hole.

What other information can we get about a black hole? Precious little, if some of the exploratory work in this field turns out to be correct! In our example of the gravitational collapse of the star, we assume the star to be a perfect sphere. The collapse of a spherical

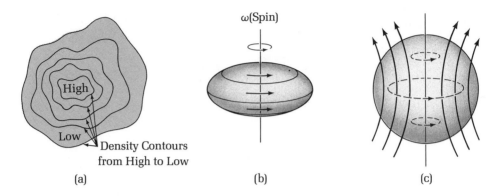

7-11. (a) Even if the collapsing body is irregular in shape, the surviving
information tells the outside observer *nothing* about these irregularities.
He can only measure its mass. (b) A spinning body has an angular
momentum. This can be determined by the outside observer even when
the body has become a black hole. (c) Even though the body may have
electric currents (*dashed lines*) and magnetic fields (*solid lines*), these
details are lost. The outside observer only measures the net electric
charge in the body. (From *The Physics–Astronomy Frontier* by F. Hoyle
and J. V. Narlikar. Copyright © 1980. W. H. Freeman and Company.)

object can be described by Einstein's equations, which have
explicit solutions. However, if the collapsing object has a highly
irregular shape, if it is rotating as well as collapsing, or if it has
electric charges and currents with their associated magnetic fields,
then the problem of solving the equations of collapse becomes
impossibly difficult. That is, given the initial stages of such a
collapse, we cannot find the detailed sequence of stages through
which the object in question becomes a black hole.

The exploratory work referred to above describes a somewhat
limited range of initial stages. It is concerned with situations
where the departure from the spherical state is rather small. The
conclusion based on such work, due largely to Richard Price, may
be described as follows.

In Figure 7-11, we see the initial stages of collapse of an irregu-
larly shaped object. All the different types of irregularities that
we referred to earlier are present. Yet, in the final stage, when the
object has become a black hole, what information is left for the
outside observer *A*? To *A*, the black hole will appear to have
a mass, some angular momentum, and some electric charge. By

designing suitable experiments, *A* can measure these three quantities but no more! Information about all other aspects that were present in the initial stages have disappeared. John Wheeler has described this situation by his oft-quoted remark "The black hole has no hair!"

Because so little information remains about the black hole to the outside observer, the detection of a black hole has to depend largely on indirect evidence. This has not deterred theoreticians, however, from putting forward ingenious scenarios in which black holes have important roles to play. We will concern ourselves with a few examples in the following chapter.

Spacetime singularity

We end this chapter with a brief discussion of what happens to *B* after he has crossed the event horizon. We must remind ourselves that, although *A* has to wait *forever* to witness *B*'s crossing of the horizon, *B* himself does not notice this slow passage of time. By his clock, the crossing of the event horizon takes very little time. In fact, by *B*'s clock, the time taken for the Sun to undergo a gravitational collapse (in the unlikely event of a sudden withdrawal of its internal pressures) is only about 29 minutes, right from its present size to the final pointlike state. The collapse time of a star from initial neutron densities is even smaller, on the order of one ten-thousandth of a second, as measured by *B*.

There are, however, other effects that *B* will begin to notice as he collapses with the star—effects that will prove to be highly uncomfortable. In Chapter 6, we described the tidal forces. These forces (see Figure 7-12) will tend to stretch *B* in the radial direction. How big are these effects? For a star three times as massive as the Sun, the tidal tension becomes 100 times the atmospheric pressure on the Earth's surface by the time the star's radius is 40 times the Schwarzschild radius. No human being can withstand this disruptive tension. And by the time the Schwarzschild radius is reached, this tension will have increased 64,000 times!

Even if *B* somehow summons incredible resources to survive tidal disruption, he has a worse fate in store. As the star contracts inside the event horizon, the geometry of spacetime around *B* will become increasingly peculiar. (The growing tidal force is just one

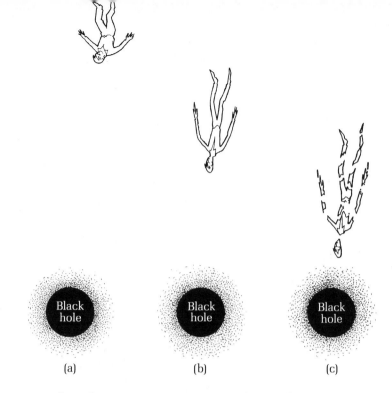

7-12. The unfortunate experience in store for *B* as he is torn apart by the huge tidal forces of the collapsing object is the same as that of a man falling freely toward a black hole. In (a) the man is far away from the black hole and hardly notices the black hole's tidal force. In (b) he is near enough to be stretched by the tidal force, while in (c) as he approaches still closer to the black hole he is torn apart by the tidal force.

manifestation of this peculiarity.) The climax is reached when the star shrinks to a point. This is "The End" for the star as well as for *B*.

For, not only does the star have infinite density, but the spacetime around it becomes infinitely curved. Mathematicians describe this as a state of *singularity*. All mathematical equations break down at this state. The laws of physics cease to operate at the instant of singularity. Beyond this instant, the future is unpredictable. To all intents and purposes, we may consider this singular epoch to mark the end of the career of the star as well as of *B*.

> *The large dark object at the center of this photograph is the giant star HDE 226868 in the constellation Cygnus. Orbiting this star is a compact object of 8 solar masses that accretes matter from the star's outer layers and emits X rays and radio waves. This object, Cygnus X–1, is presently the best-known candidate for a black hole. (Courtesy of Jerome Kristian, Mt. Wilson and Las Campanas Observatories, Carnegie Institution of Washington.)*

8

Black Holes as
Cosmic Energy Machines

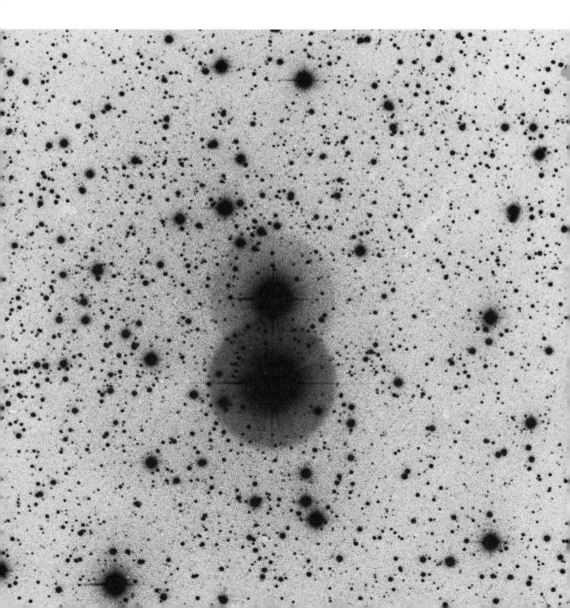

8

Black Holes as
Cosmic Energy Machines

The physics of black holes

The present literature on relativity and astrophysics reflects the popularity of the black hole as a tool of the theoretician. In the past 10 to 15 years, considerable research has been undertaken toward understanding the peculiar nature of the spacetime geometry associated with black holes. Some of this work, in the areas of differential geometry and topology, is abstract. At the same time, black holes have proved attractive to model builders trying to explain diverse astronomical phenomena, ranging from the deep interior of the Sun to very powerful sources of energy situated both within our Galaxy and beyond it. In this chapter we will try to capture the flavor of this exciting work.

We begin with a discussion of the physical properties of black holes once they have formed. In Chapter 7 we explained how black holes could form as a result of the gravitational collapse of a massive star. Later in this chapter, we will encounter other variations of this basic theme. For the present, we will assume that there are black holes in the Universe and we will ask how they interact with other matter or with other black holes. By attempting to answer such questions, theoreticians have arrived at certain basic laws of black-hole physics—laws that govern the physical behavior of black holes.

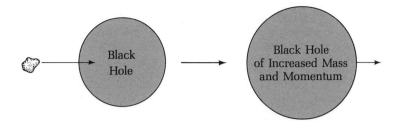

8-1. When a piece of matter (shown in the figure) falls into a black hole, the energy and momentum of the system is unchanged. This means that the black hole, after "gobbling up" the piece of matter, has increased its mass and has also acquired a momentum in the direction in which the piece was moving. This outcome is stated in the first law of black-hole physics.

The *first law of black-hole physics* states that a black hole is subject to the same conservation laws of energy, momentum, and angular momentum that control the dynamical behavior of ordinary matter (see Chapter 1). For an example of how this law operates, consider what happens when a piece of matter collides with a black hole. We have seen that a black hole does not allow even light to escape. Therefore, any piece of matter, once it crosses the event horizon (as in Figure 8-1), is "gobbled up" by the black hole. This gobbling-up process results in the loss of matter, as far as an outside observer is concerned (like the observer *A* of Chapter 7). The law of conservation of energy, however, demands that whatever energy was lost in the gobbling-up process should show up as an increase in the mass *M* of the black hole. In Chapter 7 we saw that the outside observer can in principle measure *M* and can therefore verify the first law of black-hole physics.

The *second law of black-hole physics* is potentially more revealing about the nature of black holes. To understand its significance, let us go back to our example of the black hole discussed in the last chapter. A spherical black hole of mass *M* has a spherical event horizon of radius

$$R_s = \frac{2GM}{c^2}$$

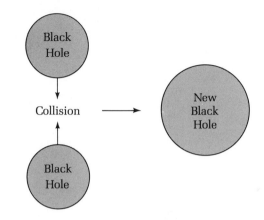

8-2. An example of the second law of black-hole physics.
Two identical spherical black holes collide and coalesce into
a single black hole. The area of the new black hole cannot be
less than the combined areas of the two original black holes.
In mathematical terms, this statement implies that the
Schwarzschild radius of the new black hole must be at least
~41.5 percent greater than the Schwarzschild radius of each
of the old black holes.

The surface area of a sphere of radius R_s in Schwarzschild's
geometry is $4\pi R_s^2$. Therefore, our black hole can be said to have a
surface area

$$A = \frac{16\pi G^2}{c^4} M^2$$

As the black hole gobbles up more matter, its mass M increases,
and therefore its surface area A also increases.

The second law of black-hole physics states that in *no* interaction of a black hole with other matter can the surface area of the
black hole ever *decrease*. If two black holes collide, they form a
larger black hole (see Figure 8-2). The surface area of the new black
hole cannot be less than the sum of the surface areas of the two
constituent black holes.

The spherical black hole described above, however, does not
bring out the full content of the second law. To appreciate the real
significance of this law, we need to look at *rotating* black holes.

The Kerr black hole

While discussing the gravitational collapse of a star, we assumed that it was always spherical during its contraction. The black hole that formed out of such a contraction is also spherical. The geometry outside the spherical black hole is that of Schwarzschild. For this reason, it is called the *Schwarzschild black hole*. It is entirely characterized by its mass M.

It may happen, however, that the collapsing star is spinning about an axis. The laws of dynamics then tell us that the angular momentum of the star must remain constant as it collapses. As a result of this rule, the star will spin faster and faster as it contracts. It may also happen that this process could lead to a disruption of the star and altogether prevent its reaching the black-hole stage. However, if this disruption does not take place, then the resulting black hole will also possess angular momentum.

The detailed process of how—if at all—a spinning object attains black-hole status is not well understood. Therefore, we cannot say definitely under which conditions a spinning black hole is formed. Nevertheless, we can say what a spinning black hole is like once it has formed. In 1963, Roy Kerr gave a description of the geometry of spacetime in the empty region outside a spinning object symmetrically shaped about its axis of spin. Kerr's solution of the problem has greatly added to our understanding of the spinning black hole, which is sometimes referred to as the *Kerr black hole*.

Figure 8-3 shows two sections of a Kerr black hole. In 8-3a we have a meridian section, that is, a section through the axis of rotation. The inner circle (solid curve) shows a section of the spherical *event horizon*. The outer broken curve denotes the section of the boundary of what is known as the *ergosphere*. We will presently see the reason for this name. In 8-3b we see a section of the black hole at a given latitude. The inner and outer circles, whose common center lies on the axis of rotation, denote respectively the sections of the event horizon and the ergosphere.

Imagine two observers B_1 and B_2. As shown in Figure 8-3b, B_2 lies outside the ergosphere while B_1 lies inside it but outside the event horizon. Now, our statement that the black hole spins about an axis has meaning only in relation to some background—say, the background provided by the distant stars. Suppose B_1 and B_2, who

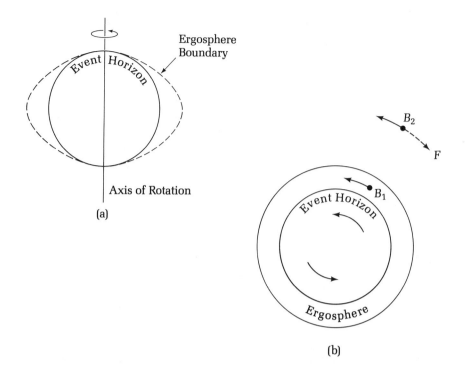

(a)

(b)

8-3. The meridian and latitude sections of a Kerr black hole are shown in (a) and (b), respectively. The two observers B_1 and B_2 are trying to remain stationary relative to distant stars in (b). Both observers have the tendency to be carried away by the rotation of the black hole in the direction of continuous arrows. B_2, who lies outside the ergosphere, can successfully counter this tendency by applying a strong enough opposing force F. An observer like B_1, who lies within the ergosphere, cannot do this. The outer boundary of the ergosphere, known as the *static limit*, is the nearest B_2 can approach the rotating black hole without being carried along by its rotation.

can see this background, want to maintain themselves at rest relative to this background. Can they achieve this?

A rough analogy of this situation is seen in Figure 8-4, in the example of an aircraft flying above the rotating Earth. An aircraft that flies above the surface of the Earth is carried along the direction in which the Earth rotates, that is, from west to east. If this were not so, it would be a simple matter to travel west: The aircraft would go up, stay in one place, and come down when the destination appeared below.

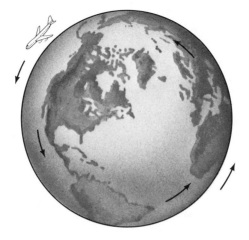

8-4. An aircraft—even in midair—is carried along in the direction of Earth's rotation.

In the case of the Kerr black hole, both B_1 and B_2 are carried along the direction in which the black hole rotates. To stay in one place, both B_1 and B_2 would have to apply extra force—say, by firing rockets—to counter this tendency of the black hole to carry them along the direction shown by the arrows. As long as an observer like B_2 is *outside* the ergosphere, stationarity can be achieved in this way. However, an observer like B_1 *inside* the ergosphere cannot remain stationary relative to distant stars *no matter how powerful the rockets are* that are employed. The observer is inexorably carried along by the spinning black hole.

What is the surface area of the Kerr black hole? As in the case of the Schwarzschild black hole, we are concerned here with the area of the event horizon. The answer for a Kerr black hole is given by the following mathematical expression:

$$A = \frac{8\pi G^2 M^2}{c^4} \left[1 + \sqrt{1 - \frac{S^2 c^2}{G^2 M^2}} \right]$$

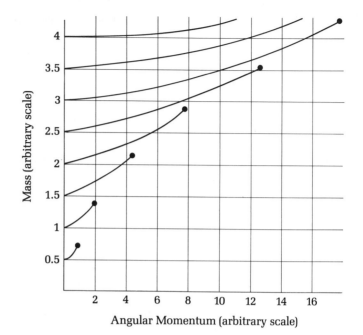

8-5. The constant area curves for Kerr black holes. Moving along any curve from right to left means passing through a succession of black-hole states with decreasing mass and decreasing angular momentum but with the same area. On the left, all curves terminate on the mass axis. These final states describe *nonrotating* black holes, that is, the Schwarzschild black holes. The second law of black-hole physics allows movement along any given curve or shifting to an upper curve from a lower curve *but not vice versa.*

None of the curves extends beyond certain fixed points on the right. These represent states of *extreme* Kerr black holes. Although the second law permits movement along a constant area curve in either direction, there is a *third* law of black-hole physics which states that we cannot attain this extreme right end of the curve with a finite series of physical operations. There is an analogy here with temperature that a physicist often makes: "The temperature of absolute zero can never be attained through a finite series of physical operations."

Notice that the area now depends on two properties of the black hole, its mass *M* and its angular momentum *S.* In Figure 8-5 we see the *constant area curves* for spinning black holes. As we move along any curve from right to left, we go through a succession of black holes with decreasing mass and decreasing angular momentum. The second law of black-hole physics allows a transition to be made from a lower curve to an upper curve *but not vice versa.*

Keeping in mind this limitation imposed by the second law of black-hole physics, we note that, no matter what physical process we use, we cannot decrease the area of a black hole, but we can at best hope to keep it constant. In that best case, we can change a black hole through a succession of states represented by any of the curves. If we go from right to left, the black hole loses mass and angular momentum; if we go from left to right, the reverse happens. Herein lies a clue to the extraction of energy from a spinning black hole.

The Penrose mechanism

Roger Penrose, the English relativist, suggested a mechanism by which energy can be extracted from a spinning black hole. The trick lies in taking advantage of the $A = constant$ curves in Figure 8-5, where we saw that, by moving along such a curve from right to left, we encountered the successive stages of a black hole —stages in which its mass and angular momentum decrease.

In the Penrose mechanism, the mass and the angular momentum of the black hole are *reduced* by the process described below. The quantity by which the mass of the black hole is reduced becomes available as energy according to Einstein's formula $E = Mc^2$ (see Chapter 5).

The principle behind the Penrose mechanism is shown in Figure 8-6. A chunk of matter is fired at the spinning black hole with sufficient energy and angular momentum so that it enters the ergosphere and circles around the axis of rotation of the black hole. At some stage, the chunk is designed to break up into two pieces. One piece falls through the event horizon and is lost forever. The other piece, however, acquires sufficient kinetic energy to escape from the ergosphere back to outer space.

Penrose showed that it is possible to adjust the parameters of this process so that the total energy of the surviving piece as it emerges from the ergosphere is *greater* than the energy of the entire chunk as it went in! In this process, the surviving piece acquired part of the angular momentum and energy of the black hole. After the process is over, the black hole will have gone into a state of lower

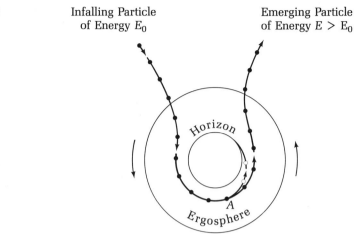

Infalling Particle
of Energy E_0

Emerging Particle
of Energy $E > E_0$

8-6. A schematic diagram describing the Penrose mechanism of
energy extraction from a rotating black hole. (From *The
Physics–Astronomy Frontier* by F. Hoyle and J. V. Narlikar.
Copyright © 1980. W. H. Freeman and Company.)

mass and lower angular momentum than before. Thus, it is possible in principle for an advanced civilization to operate the spinning black hole as a powerhouse by simply firing chunks of matter
at the black hole and receiving back pieces with greater energy
than the energy spent on the in-going chunks.

Yet the process cannot go on forever. Figure 8-5 sets the limitations on the process. If operated with perfect efficiency, the black-
hole area remains constant. However, any $A = constant$ curve
terminates when it hits the mass axis—that is, when the black hole
has lost all its angular momentum. The black hole in this state is
none other than the spherically symmetric Schwarzschild black
hole that was introduced in Chapter 7. This nonspinning state
represents an *irreducible* state of the black hole—that is, a state
from which there can be no further energy extraction. The mass of
the black hole in this state is called its *irreducible mass.*

If the Penrose process is *not* operated with perfect efficiency, it
would result in an *increase* in the area of the black hole. In Figure
8-5, we would have to jump from one curve to another that corresponds to a higher value of A. If this curve-jumping happens, the
final irreducible state is reached sooner and with a higher irreducible mass than if the area had remained constant throughout. But

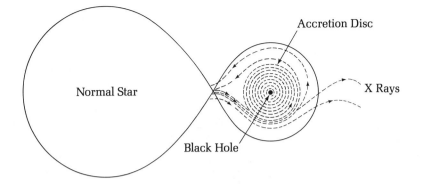

8-7. The formation of an accretion disc around a black hole in a binary star system. (From *The Physics–Astronomy Frontier* by F. Hoyle and J. V. Narlikar. Copyright © 1980. W. H. Freeman and Company.)

we have to pay for this inefficiency; we are able to extract less energy from the black hole than we could have done had we been able to maintain the area of the black hole constant throughout.

Ingenious though the Penrose process is, it is highly esoteric. So far, no astronomers have been able to "cook up" a scenario based on the Penrose "recipe." There are, however, plenty of examples in high-energy astrophysics where rotating black holes have been suggested as sources of energy. In the examples discussed below, it will become apparent that the main property of black holes utilized in the production of energy is their concentration of a lot of matter in a small region of space.

Cygnus X–1

In Chapter 6, we described the tidal effects that arise in a binary star system. Let us follow that scenario further.

Recall that, in the binary system shown in Figure 6-8, two stars A and B go around each other in elliptical orbits. The star B is compact, while the star A is extended. Tidal effects become important when A extends beyond its *Roche lobe*. The surface matter from A is then pulled by its companion B. What happens to this pulled-out material?

Figure 8-7 illustrates what happens in a schematic way. If the double star system were not rotating, the material pulled out from

A would have gone straight toward *B*. Because of rotation, however, this material does not go straight away to *B* but is made to rotate around *B*. As shown in Figure 8-7, the matter from *A* goes around and around *B* mostly in their common equatorial plane, where the tidal effect is largest. It finally falls into *B*. This process results in the formation of a disc of infalling matter around *B*.

In the 1940s, Hermann Bondi, Raymond Lyttleton, and Fred Hoyle considered various situations where a star can *accrete* matter from interstellar space as it moves through it. The strong gravitational pull that the star exerts on the surrounding matter explains the accretion. The problems considered by these astrophysicists were largely of spherical accretion—that is, accretion from all directions onto a spherical star. Bondi's classic paper of 1951 formed the starting point of modern accretion theories.

We now consider the same concept of accretion in the context of a binary star system. The difference here is that the accretion is in a disc rather than from all directions. The size of the disc is determined by various parameters, such as the rate of accretion, the rate of infall into star *B*, the mass of star *B*, and so forth. The size of the disc can fluctuate if the accretion rate varies. The interesting effect of such a disc is, however, that the matter in it becomes *hot* as a result of friction, and this hot matter radiates. The radiation frequency depends, among other things, on the temperature of the disc. Calculations show that this radiation should be largely in the form of X rays.

The main role in this picture is, of course, played by the star *B*. For a radiating accretion disc to form around it, the star *B* must be a very compact star. In the early 1960s, neutron stars were discussed anew by astrophysicists; in the late 1960s, black holes began to gain prominence. Thus, both neutron stars and black holes were recognized as good candidates for star *B* in this accretion-disc mechanism.

Along with this theoretical development, the progress of space technology made it possible for astronomers to observe cosmic X-ray sources. Since X rays from outer space are absorbed by the Earth's atmosphere, their observation becomes feasible only if we can set up detector instruments above the atmosphere. Such

attempts were made in the 1960s. A major advance in X-ray astronomy was the survey conducted by the *UHURU* satellite* launched
on December 12, 1970. This survey revealed the existence of many
X-ray sources, among which was one in the constellation of
Cygnus. This source, now known as Cygnus X–1, is the best
example of the binary star X-ray emission described above.

The binary stary system at the location of Cygnus X–1 consists of
a supergiant star *A* and a compact companion *B*. The star *A*,
known as HDE 226868 in the Henry Draper Catalogue, is the only
visible star of this binary system. Its mass is at least 20 $M_\odot$, and it
appears to be orbiting about its companion, taking about $5\frac{1}{2}$ days
to complete one orbit. The companion *B* cannot be seen, but its
existence can be inferred. It is this object that has caused so much
excitement among black-hole astrophysicists.

Detailed observations of the orbital parameters of this double-
star system suggest that the mass of *B* must be at least 8 $M_\odot$. This
high value far exceeds the theoretical upper limit on the mass of a
neutron star, a limit that is known not to exceed 3 $M_\odot$ and is probably closer to 2 $M_\odot$. This result therefore eliminates *B* as a standard
neutron star. What else can it be? The alternative is a black hole.
For this reason, Cygnus X–1 is often cited as the best-known observational evidence for the existence of a black hole.

While most astronomers accept this conclusion, dissenting
astronomers and astrophysicists point out that our knowledge of
the state of matter at very high densities (in excess of a million
billion times the density of water) is still in a highly speculative
stage. Additional information may lead us to revise the upper limit
on the mass of a compact, dense star. Moreover, most other X-ray
sources in binary stars are consistent with the supposition that star
B could be a neutron star. Thus, Cygnus X–1 seems more of an
exception than a rule. (Circinus X–1 is another but weaker case for
a black hole.) Could it be that more than two stars are involved in
the system that gave rise to Cygnus X–1, with none of the compact
components exceeding the mass limit for a neutron star? There are

---

*UHURU means "freedom" in Swahili, the language of Kenya, where the satellite was launched on Kenya's Independence Day.

hints of additional periodicities in this system that could arise in a three-star complex.

However, if the black-hole interpretation of Cygnus X–1 is speculative, the other current alternative interpretations are even more so. In any case, the example of Cygnus X–1 has prompted many black-hole astrophysicists to be more daring and to consider the applications of the accretion-disc scenarios on even larger scales. We consider next a few examples of such applications.

Supermassive black holes

Considerations of stellar evolution led us to the concept of black holes with masses that are a few times the mass of the Sun. The black hole believed to exist in the double star system giving rise to Cygnus X–1 has a mass of about 8 $M_\odot$. We now consider black holes far more massive than those arising as the end states of stars. These black holes, known commonly as *supermassive black holes,* are thousands to billions times more massive than the Sun, and they are conjectured to exist in globular clusters, galactic nuclei, and quasars. Their role in these astronomical systems is to generate energy.

Figure 8-8 shows a globular cluster, and Figure 8-9 shows the galaxy M 87 (M refers to the *Messier Catalogue*). A globular cluster is a cluster of hundreds of thousands of stars moving under the gravitational influence of one another. Our Galaxy contains many globular clusters, and the one shown in Figure 8-8 is a typical example. Notice how the concentration of stars is largest in the center and thins out toward the periphery of the cluster. It is believed—though as yet no detailed theory exists to substantiate this belief—that the dense concentration of matter in the center of a globular cluster leads to the formation of a black hole. The mass of a black hole in a globular cluster may be as high as 1000 $M_\odot$.

Supermassive black holes in the nuclei of galaxies like M 87 are even more massive. The mass of the black hole inside M 87 is believed to be as high as 5 × 10^9 $M_\odot$! Speculations apart, what evidence is there in support of such claims?

8-8. Photograph of the globular cluster NGC 6624. (From M. H. Liller and B. W. Carney, *The Astrophysical Journal* 224, 385. Courtesy of M. H. Liller.)

8-9. Photograph of the galaxy M 87. Notice the jet coming out from the center. Careful studies of the central region of this galaxy suggest that there may be a black hole 5 billion times as massive as the Sun in the nucleus of the galaxy. (Palomar Observatory photograph.)

First, let us try to understand what these supermassive black holes are expected to *do*. In general, if such black holes do form, they must arise from the gravitational collapse of a collection of stars. Such a collection of stars has an overall angular momentum that survives even when the black hole is formed. The black hole therefore is of the spinning (Kerr) type. And the spinning black hole tends not only to attract the surrounding matter but also to move this matter around the axis of rotation. The effect is largest in the equatorial plane of the black hole, where this surrounding matter forms an accretion disc. This accretion disc is qualitatively similar to that which forms around a black hole in a binary system, although in actual size and mass it is enormously bigger. The black hole may generate energy through heating this disc as in the binary system.

There are X–ray sources in globular clusters. For example, the globular cluster shown in Figure 8-8 is known to contain an X-ray source called 3 U 1820–30 (U refers to the *UHURU Catalogue*). The disc surrounding a black hole may well account for such globular-cluster X-ray sources.

With regard to M 87, two teams of astronomers using different observational techniques came to the conclusion in 1978 that a black hole about $5 \times 10^9 \, M_\odot$ in the nucleus of the galaxy may best account for their observations. One team of astronomers from the Hale Observatory and the Jet Propulsion Laboratory measured the brightness of visual light from all over the galaxy. Although such brightness measurements had been made previously for many galaxies, the sensitivity achieved in the 1978 results was far higher than before, which enabled the astronomers to look at the nuclear region of M 87 more carefully. They found that the brightness profile of the galaxy rose sharply toward the center instead of flattening out, as shown in Figure 8-10. Such a rise in luminosity indicates a dense conglomeration of stars in the nuclear region, which in turn points to a strong gravitating object that keeps the stars huddled together near the center. What could this object be? After examining several alternatives, the observers found the black hole interpretation to be the most suitable in this context.

The other team of astronomers (from the Hale Observatory, the Kitt Peak National Observatory, the University College, London,

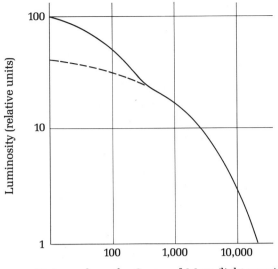

Distance from the Center of M 87 (light-years)

8-10. The luminosity profile of M 87. Notice that, instead of reaching a flat plateau toward the center (as shown by the dashed curve), the luminosity shows a sharp rise, which points to some compact energy source in the center of M 87. A favorite explanation for this energy source is that it is a black hole. The graph uses a logarithmic scale both for luminosity and distance. The rise in luminosity from the outer end of the profile toward the peak at the center is by a factor of 100.

and the University of Victoria) examined the nuclear region spectroscopically. Their measurements give indications of the extent of spread in the speeds of stars in the nuclear region. The larger-than-expected spread in speeds again led these astronomers to conclude that a supermassive black hole in the nucleus caused this spread.

Do black holes exist?

More recently, the supermassive black hole theory has been applied to other areas in astronomy where vast quantities of energy seem to come from small volumes. Quasars and the central regions of radio galaxies are examples of such applications.

Just how firm is the actual support of the highly speculative theories we have been discussing in this chapter and in Chapter 7?

There are, of course, many leading astronomers who feel that the credence given to black holes is somewhat exaggerated. Since black holes cannot be directly observed (by definition!), their existence must be *inferred* from indirect measurements. These examples illustrate the maxim of Sherlock Holmes, the great fictional detective of Sir Arthur Conan Doyle, which may be paraphrased as follows: "When all other plausible astrophysical explanations fail, the black hole interpretation, however bizarre it may seem, must represent the truth."

Some astronomers feel that all other plausible astrophysical explanations have *not* been fully explored, and therefore the inference of the existence of black holes rests on insufficient evidence. This skepticism about astronomical black holes, which arises in part from their bizarre nature, again reminds us of the Black Hole of Calcutta. The description of the Black Hole of Calcutta is so macabre that a few historians doubt whether the event took place at all!

Particle-antiparticle production in the laboratory. A high-energy gamma ray collides with an electron (top, center), producing an energetic recoil electron (left center) and an electron-positron pair (top left, right). A second gamma ray produces a second electron-positron pair (center). (Courtesy of Lawrence Berkeley Laboratory, University of California.)

9

White Holes:
Myth or Reality?

9

White Holes:
Myth or Reality?

Can black holes be white?

Until 1974, physicists believed that a black hole is the ultimate in darkness. The event horizon of all black holes presents a barrier to knowing what may be happening inside. Even the spacetime singularity inside the event horizon (discussed in Chapter 7) is hidden from the outside observer. A cosmic censorship seems to operate so that the event horizon prevents any information from leaking out.

These ideas were rudely shattered in early 1974 by a remarkable result derived by the Cambridge theoretician Stephen Hawking. After a complicated series of calculations, Hawking concluded that *a black hole radiates!* For each type of black hole, whether spinning or nonspinning, it is possible to associate a "temperature," and if the black hole happens to be situated within a surrounding region of lower temperature, it would radiate energy.

What was different about Hawking's calculations that led him to such a remarkably different conclusion? Let us try to gain a qualitative understanding of the differences involved.

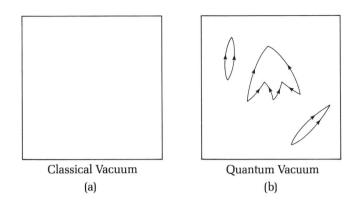

Classical Vacuum   Quantum Vacuum

(a)   (b)

9-1. The spacetime diagrams for (a) the classical vacuum and (b) the quantum vacuum. In (a), *nothing* happens. In (b), particle and antiparticle pairs are being created and annihilated all the time. Note that *no* particle or antiparticle survives for long, and the net effect of these fluctuations at any time is zero.

In Figure 9-1a, we see the spacetime diagram for an empty region, or *vacuum*. As expected, there is nothing there! But this expectation of nothingness is based on the laws of motion and interactions of classical physics, which are applicable to the *macroscopic* world around us. It was by observing the macroscopic world that Galileo and Newton formulated the concepts of motion and such physicists as Coulomb, Ampere, Faraday, and Maxwell arrived at a comprehensive picture of electric and magnetic interactions. By the end of the nineteenth century, however, *microscopic* observations—observations of the properties of atoms and molecules—were beginning to reveal cracks in this classical framework. The new rules, now known as the *quantum theory*, endow the microworld with a far richer structure than could ever have been imagined on the basis of observations of the macroworld only. For example, Figure 9-1b shows what a vacuum is like in quantum theory. Far from being empty, it is a jumble of *virtual* particles and antiparticles. A virtual particle or antiparticle does not have a permanent existence—a particle and antiparticle pair annihilate shortly after creation. As shown in Figure 9-1b,

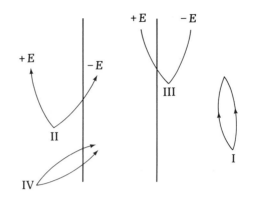

9-2. The thick vertical lines denote the boundary of a black hole. The four cases of likely outcome of creation of particle–antiparticle pairs in a vacuum are shown by I, II, III, and IV. The laws of quantum mechanics in the curved spacetime of a black hole determine the net outcome of these cases.

such particle–antiparticle creation and annihilation are constantly going on within the vacuum. Thus, even though the average overall effect is zero, as in the classical case of Figure 9-1a, the fluctuations from the *zero state* are considerable and cannot be ignored. Their effect has been observed in several experiments in atomic physics.

These quantum fluctuations play a vital role in the Hawking process. In Figure 9-2, we see a black hole in a quantum vacuum. The two straight lines represent the boundary of the event horizon in a spacetime diagram. Again we have quantum fluctuations. In Case I shown in Figure 9-2, a pair of particles and antiparticles is created and then annihilated as before. The pair is apparently unaffected by the black hole. In Case II, however, one member of the pair is attracted by the black hole and falls through the event horizon. Its companion escapes and eventually reaches regions remote from the black hole.

How should we interpret Case II? To an outside observer located far away from the black hole, the observable aspect of Case II is the appearance of a particle apparently produced by the black hole!

In a virtual pair, if one member of the pair has energy E, the companion has energy $-E$ (the total energy for the pair being zero). If the member with energy E has escaped, the black hole has, in this process, generated energy E.

What has happened in this process to the energy of the black hole? Because it gobbled up an object of energy $-E$, it will now have its total energy *reduced* by the quantity E. Thus, the overall effect seems to be that the black hole has produced a particle of energy E and at the same time lost an amount E from its own energy reservoir.

It could also happen that a particle of energy E is gobbled up by the black hole, and its companion of energy $-E$ escapes. In this process—let us call it Case III—the black hole gains energy. We can also think of another process (shown by Case IV) where both members of the pair fall into the black hole.

Hawking's calculations involved taking account of the rates at which the various processes could occur. His conclusion was that the overall effect is of Case II—that is, a black hole situated in a vacuum has a tendency to generate energetic particles.

The quantitative calculations are intricate and involve issues that are not yet fully resolved. Yet experts in the field agree that the effect predicted by Hawking should exist. Let us therefore try to understand the implications of this effect for the physics of black holes.

First, unlike the Penrose process for energy extraction from black holes (see Chapter 8), the Hawking process works for *all* types of black holes. For the simplest—the Schwarzschild black hole—the radiation of energy from a black hole of mass M is the same as that expected from a body heated to a steady temperature of 6×10^{-8} times the number of solar masses in the black hole, or

$$T = 6 \times 10^{-8} \frac{M_\odot}{M} \text{ degrees Kelvin}$$

This temperature is measured on the absolute (Kelvin) scale in the above formula. Zero on the absolute scale is approximately -273 degrees on the centigrade scale, and one-degree increases on the absolute scale equal one-degree increases on the centigrade scale. For example, water boils (at sea level) at 100°C. This temperature equals 373 degrees on the absolute scale.

Just how hot are black holes? We saw that stellar black holes have masses exceeding 3 $M_\odot$. Thus, a black hole of 6 $M_\odot$ will have a temperature of 10^{-8} K, that is, a hundred-millionth of a degree. At such low temperatures, the black holes would hardly radiate! Therefore, our concept of a black hole being black is not altered by Hawking's result, at least as far as stellar black holes or supermassive black holes are concerned. Since these are the only types of black holes we have so far encountered, is the Hawking process only of academic interest?

Notice that the temperature of a black hole increases as its mass decreases. If black holes of masses considerably lower than 1 $M_\odot$ existed, these black holes would be considerably hotter. For example, a black hole of mass as low as eight thousand-billionths of the mass of the Sun would have a temperature of about 7200°C. This is the temperature at which a body becomes *white hot*. Thus, a sufficiently low-mass black hole is white hot!

Can such low-mass black holes ever form? Some theoreticians have conjectured that, in the first few moments after the creation of the Universe (Chapter 10), the geometry of spacetime may have been highly nonuniform; in this early stage, the inhomogeneities of geometry could manifest themselves as tiny black holes. These *mini black holes* may have a mass as low as 10^{14} grams. Naturally, the Hawking process is very significant for them.

Figure 9-3 shows what happens to the mass of a mini black hole in the course of time. As it radiates energy, it loses mass. This reduction of mass increases the temperature of the black hole, as per the formula we have just described. The black hole therefore loses energy even faster. The process of radiation gathers speed, and in a finite time the entire mass of the black hole is "evaporated."

Can this time be long enough for such a primordial black hole to survive to the present? As we shall see in Chapter 10, the age of the Universe is about 10 billion years. How massive must a mini black hole be in order to last this period of time? The answer is, about 10^{14} grams *at least*.

It has been suggested that a primordial black hole of mass about 10^{14} grams might be in its final stages of existence, and hence its evaporation would be very fast. Can it be observed? Calculations suggest that the products of such evaporation would be particles or radiation of very high energy. Since the early 1970s, gamma-ray

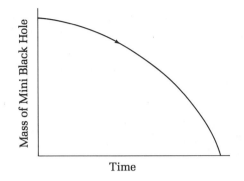

9-3. A graph showing how the mass of a mini black hole decreases with time as it radiates via the Hawking process. The rate of decrease is rapid in the final stages. (Figure not drawn to scale.)

astronomy from satellite-based detectors has revealed the existence of gamma-ray bursts. In a typical burst, gamma rays—that is, radiation of very high energy photons—emerge from a source for a very short period (on the order of 1–10 seconds). Could these bursts be due to the evaporation of primordial black holes? Detailed investigations by Bernard Carr indicate not; nevertheless, there may be other ways in which the final stages of a primordial black hole produce observable results.

White holes

Because the Hawking black holes pour out energy rapidly and would therefore be visible as very bright objects, they are sometimes referred to as white holes. This description is somewhat misleading, because the term *white hole* had been used earlier to describe another phenomenon long before the Hawking process was discovered.

These *other* white holes also pour out energy in an explosive fashion; but unlike the mini black holes, they do not rely on a quantum effect. To understand how these *classical white holes* operate, we have to recall the phenomenon of gravitational collapse leading to the formation of a black hole (discussed in Chapter 7).

In our examples of the collapsing star, we had placed one observer, B, on the collapsing surface, while the other observer, A, was placed far away from the star. The light waves sent from B to A were *redshifted* for two reasons: Part of the redshift was of gravitational origin, and part was of Doppler origin.

In a white hole, we have a *time-reversed version* of a collapsing object as seen by B. What does time-reversal mean? When we observe any phenomenon taking place in nature, we see a sequence of states (of whatever we are observing) ordered according to chronological time. A good example is a movie film, where we seem to see some event taking place. But the film in fact consists of a succession of still pictures passing rapidly in front of our eyes. If we were to reverse the succession of these still pictures, the film would run backwards. If the sequential order of the original event were reversed, the event seen on such a film would be a new type of event.

Many movie projectors have the capability of running the film backwards, and very often this process leads to hilarious new events being seen on the screen—events in which people walk backwards, food is taken out of the mouth, and the waters of Niagara Falls go up!

The events seen when a movie runs backwards are the *time-reversed* versions of the normal sequence of events. To the physicist, time-reversal has a deep significance. The basic laws of physics—whether laws of gravity, electricity, magnetism, or the interactions in the interior of the atomic nucleus—are all *time-symmetric*. This means that any event taking place according to these laws has a time-reversed version that can also take place according to the same laws. The laws *per se* do not make any distinction between the event and its time-reversed version. If these "strange" events are also possible, why do they not actually take place? It is believed that, in spite of the basic time-symmetry of physical laws, some selection process might be operating in the Universe, a process that allows some events to take place but *not* their time-reversed versions.

In Figure 9-4a, we see some of the different stages during the collapse of a massive spherical star. These correspond to the still frames of a movie film of gravitational collapse. In Figure 9-4b, the order of these stills is reversed, and we see a sequence of states

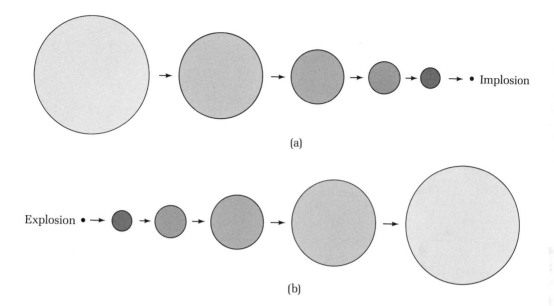

9-4. In (a) we see in chronological order a few stages in the collapse of a star to a singular state. In (b) the same stages are shown in a time-reversed sequence. The object now appears to explode into existence and expand thereafter. The time coordinate is that of an observer on the surface of the object.

of expansion. The *implosion* into a singularity in Figure 9-4a is now replaced by an *explosion* from a singularity. Notice that the singular state, which was the *final* state of collapse, has now become an *initial* state of explosion.

Since Einstein's general theory of relativity is a time-symmetric theory, it permits the explosive situation of Figure 9-4b, just as it permits the implosive situation of Figure 9-4a. This symmetry enables us to talk of white holes as the time-reversed phenomena corresponding to black holes.

Notice, however, that the symmetry of the two situations of Figure 9-4 is manifest in the frame of reference of an observer on the surface—our observer *B* of Chapter 7. To an outside observer like *A*, the black hole and white hole do not appear as time-reversed versions of each other. We will now discover the reason why.

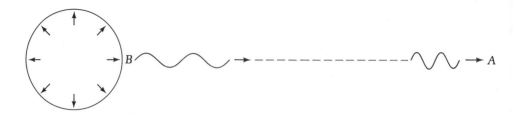

9-5. Light emitted in the radial direction from B to A can be blueshifted in the early stages of a white hole. The waves emitted by B are shown to be blueshifted by the time they reach A; their wavelength has been *reduced* to half the original value.

Blueshift from white holes

Let us see what happens to our light propagation problem in the case of a white hole. In Figure 9-5, we see a radial light wave from B to A. As in the case of collapsing objects, this light travels from a region of strong gravitational effects to a region of weak gravitational effects. Thus, had B been at rest relative to A, the light sent out from B to A would have suffered a gravitational *redshift* (as described in Chapter 7).

However, B is not at rest relative to A. The surface on which B is located is moving *toward* A; hence, the change in the wavelength of light due to the Doppler effect must be considered. The Doppler effect also operated in the collapse problem described in Chapter 7; the difference now is that the Doppler effect produces an *increase* in the frequency and a *decrease* in the wavelength of light as it travels from B to A. Applying this result to visible light, we would expect the Doppler effect to shift all wavelengths toward the *blue* end of the spectrum (since blue is the color near the short-wavelength end).

We therefore have two opposing effects—the gravitational effect tending to increase the wavelength of light from B to A (the redshift), and the Doppler effect doing the reverse (the blueshift). Which of the two effects would win out?

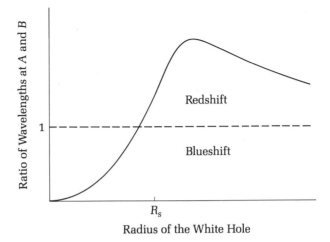

Radius of the White Hole

9-6. The ratio λ_A/λ_B is very small in the early stages of a white hole, implying the existence of large blueshifts. The ratio increases as the white-hole expansion slows down. It may become 1 either before or after the white hole bursts out of its Schwarzschild radius R_s. (In the figure, this happens before the Schwarzschild radius is reached.) Beyond this stage, the blueshift gives way to redshift of gravity. In late stages of the explosion, the ratio λ_A/λ_B begins to decline as the gravitational redshift also becomes weaker and weaker.

Figure 9-6 illustrates the answer in a qualitative fashion. Here we have plotted the ratio λ_A/λ_B, the ratio of the wavelengths received by A and emitted by B, against the size of the white hole at the time of emission. Notice that the dashed straight line corresponds to $\lambda_A = \lambda_B$. Above this line there is redshift, and below it there is blueshift. It is clear that in the early stages of the expansion, when the speed of B toward A is large, the Doppler effect wins out, and the overall effect is of blueshift. In the later stages, the expansion has slowed down and the gravitational redshift is more important than the Doppler blueshift.

Not only does light from the white-hole surface emerge from deep inside the Schwarzschild sphere $R = R_s$, it also emerges with increased energy. Thus, the event horizon that plays the role of cosmic censor in black-hole physics does not prevent the signals by B on the white-hole surface from leaking outward. To someone only accustomed to black holes, this behavior of white holes may appear peculiar, but it is nonetheless true.

In 1964, Sir Fred Hoyle, John Faulkner, and the author first suggested that this behavior of white holes makes them ideally suited for generating radiation and particles of high energy. A decade later, Krishna Apparao, Naresh Dadhich, and the author carried this work further. This chapter ends with a brief discussion of the plausibility of the role of white holes as generators of high-energy particles.

White holes as particle accelerators

Astronomers encounter several instances of sources of particles or of high-energy radiation in the cosmos. The highest energy particles in the cosmic rays that bombard the Earth's atmosphere are protons, whose energy per particle is as high as 10^{20} electron volts. To put this number in a proper perspective, we have to remember that this is about a *hundred billion* times the energy of the proton at rest. The large increase in energy arises because the proton is moving with a speed very close to the speed of light. What mechanism caused the proton to move with such high speed? So far, the conventional sources known to astronomers —like an exploding supernova (Chapter 4) or a rapidly rotating pulsar (Chapter 7)—come nowhere near producing such highly energetic particles.

Examples of radiation of high-energy photons are becoming known since the progress of X-ray and gamma-ray astronomy. Of special interest to us here are the "burst" sources, that is, sources pouring out X rays or gamma rays over very short intervals, of the order of a second or less. Any astronomical object showing changes over such a short time scale must necessarily be compact. A general rule of thumb used by the astronomer in setting a limit to the linear size L of a source is the simple formula

$$L < c\,T$$

That is, L cannot exceed the product of the speed of light c and the time scale T. For example, a source of time scale $T = 1$ second will have a size not exceeding 1 *light-second* (which equals 186,000 miles).

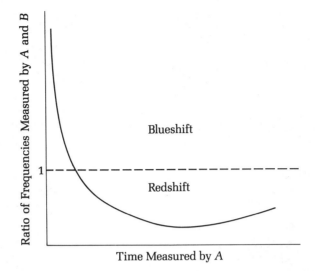

9-7. Instead of plotting the ratio of wavelengths measured by *A* and *B* against the white-hole size as in Figure 9-5, we now show how the ratio of frequencies measured by *A* and *B* changes with the time kept by *A*'s clock. The blueshifts last for a very short time in the beginning, which indicates that if a white hole is to generate high-energy particles and radiation, it will do so in a short time, in a burst. (Figure not drawn to scale.)

What could be the central mechanism that, acting over a compact region of space, produces matter and radiation of such high energies? Among the various possibilities considered by astronomers, the white hole appears to have the right requirements.

In Figure 9-6, we saw how a white hole produces very large blueshifts in light waves going *outward* from its surface. The large blueshift phase, however, lasts only while the size of the white hole is very small. In Figure 9-7, we note that this phase lasts a very short time, and therefore the radiation poured out by the white hole does have the characteristics of a *burst*. A blueshift can be high enough in the early stages to convert visual radiation to X rays or even gamma rays.

The same effect can also lift the energies of particles emitted from the surface of a white hole to the high values seen in cosmic

9-8. Photograph of the galaxy NGC 5128, identified with the radio source Centaurus A. This radio source (in common with many similar radio sources) may have originated in an explosion. (Palomer Observatory photograph.)

rays. As with radiation, this effect operates only in the early stages of the white-hole explosion.

White holes could exist in astronomical objects displaying signs of explosion. The quasars, radio galaxies, and the nuclei of Seyfert Galaxies are likely sites for white holes (see Figures 9-8 and 9-9).

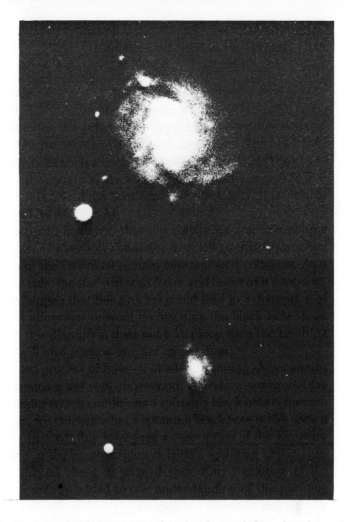

9-9. The Seyfert galaxy NGC 1068, whose nucleus is believed to show signs of explosion. The bottom half of the picture is underexposed to show only the nuclear region of the galaxy. (Indian Institute of Astrophysics.)

Can white holes exist?

In spite of their attractive possibilities in high-energy astrophysics, white holes have not gained the acceptance they deserve. In contrast to black holes, whose observable effects are necessarily of an indirect type, the effects of white holes are directly observable. What is the reason for the apparent neglect of white holes by astronomers?

There are several reasons why white holes have been poorly received. One easily understood reason is that we do not know what causes a white hole to erupt. That its initial state is of a singular character is clear. But how did this state come about in the normal course of astronomical development? For a black hole, on the other hand, a recipe for formation exists—at least in the case of stellar black holes. In Chapter 7, we saw that such black holes form toward the end of the evolution of a massive star.

In 1964, Nee'man from Israel and I. D. Novikov from the USSR independently suggested that white holes are "delayed" events representing creation of matter—delayed with respect to the epoch when the entire Universe was created some 10 billion years ago (see Chapter 10). The singular origin of a white hole is therefore a "lagging" event that has taken place comparatively recently.

Recent work by D. M. Eardley in the United States and by K. Lake and R. C. Roeder in Canada seems to considerably restrict this *lagging core* scenario for white holes. The difficulty is this: In the early stages, as the white hole explodes, its outer surface may encounter ambient matter. Relative to the white-hole surface, this matter moves inward, and even a small density of this matter can smother the tendency of the white-hole surface to move outward. The result is a slowing down of the outward motion and the ultimate collapse of the object into a black hole. These investigations therefore concluded that white holes are highly unstable objects, except when they occur soon after the origin of the Universe. As such, they will not be of much use in explaining more recent astronomical phenomena.

However, it seems premature to pass judgment on white holes on the basis of these objections alone. Their investigations have been based on extrapolations of our concepts of stability to situations when the geometry of spacetime was very much different than it is now. Questions remain even in the theory of black holes, although their role in astrophysics has been investigated in great detail. White holes also deserve further investigation.

In spite of these questions regarding white holes of a limited size, astronomers do not hesitate to discuss the biggest white hole of them all. As we shall see in Chapter 10, current theories hold that the Universe itself is a gigantic unbounded white hole!

A map of galaxies in the northern galactic hemisphere, showing the large-scale structure of the Universe. The map includes all galaxies down to nineteenth magnitude. Although many clusters of galaxies are apparent, on the largest scale the distribution of galaxies appears uniform. (Courtesy of P. J. E. Peebles.)

10

The Expanding Universe

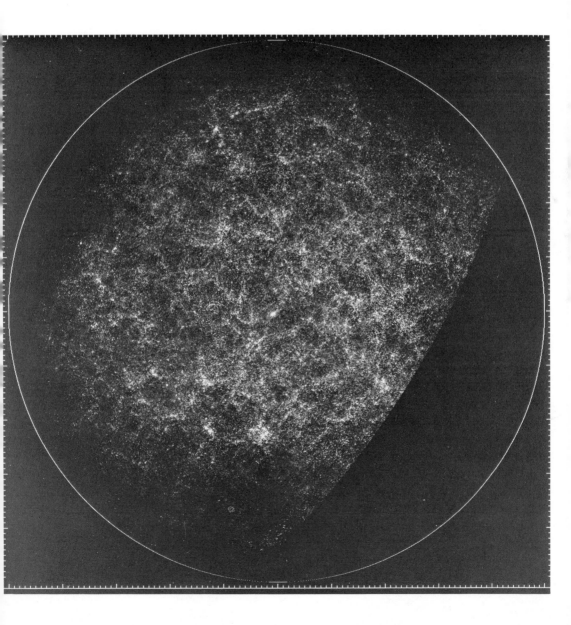

10

The Expanding Universe

The static Universe in Newton's theory

Back in the 1690s, Isaac Newton attempted an ambitious application of his law of gravitation. Newton wanted to describe, with the help of his theory of gravity, the largest physical system that can be imagined—the Universe. How did Newton fare in this attempt?

In a letter to Richard Bentley dated December 10, 1692, Newton expressed his difficulties in the following words:

> It seems to me, that if the matter of our Sun and Planets and all ye matter in the Universe was evenly scattered throughout all the heavens, and every particle had an innate gravity towards all the rest and the whole space throughout which this matter was scattered was but finite: the matter on ye outside of this space would by its gravity tend towards all ye matter on the inside and by consequence fall down to ye middle of the whole space and there compose one great spherical mass. But if the matter was evenly diffused through an infinite space, it would never convene into one mass.

Figure 10-1, which illustrates a finite and uniform distribution of matter in the form of a sphere initially at rest, helps explain Newton's difficulty. Will the sphere stay at rest forever? The matter in the sphere has its own force of gravity, which tends to pull the different parts of the sphere toward one another, with the result that the sphere as a whole contracts. We have encountered this force of self-gravity in stars (Chapter 4) and in the phenomenon of

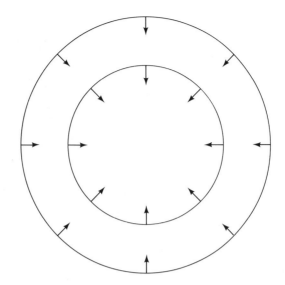

10-1. A finite spherical object tends to contract if there is no pressure to withstand gravity. This figure shows the boundary sphere and a typical concentric sphere inside the boundary sphere. The arrows indicate the direction of contraction.

black-hole formation (Chapter 7). In the case of stars, the internal pressures oppose self-gravity and maintain the stars in a static shape. In the formation of a black hole, these pressures are negligible compared to the self-gravity, with the result that a gravitational collapse of the object ensues. What happens to the matter distribution in Figure 10-1?

Because of our assumption of uniformity, the pressure forces do not operate in this system at all! So the sphere undergoes a gravitational collapse: *It cannot remain static.* Not only would the sphere shrink, it would also attract any bits of matter located outside, as Newton noted in his letter.

There is a difference, however, between our *finite* system of Figure 10-1 and an *infinite* Universe. The finite sphere collapses toward its center. Where should the Universe collapse to? A little consideration will show that there is *no central position* in a uniform infinite Universe. So there is no *net* tendency to collapse!

To put it differently, consider a typical point P in our sphere of Figure 10-1. If we take into account the overall pull of attraction of all points in the sphere, we find that P is attracted toward the center of the sphere (in the direction shown by the arrows). However, a typical point P in an infinite Universe is pulled equally in all directions and hence stays where it is.

Although theoretically a static Universe is possible from this reasoning, Newton realized that it would be highly unstable. A small departure from uniformity would lead to an *enhancement* of that departure. As a result, the Universe would break up into finite-sized clumps of matter that would undergo gravitational collapse, as did our sphere of Figure 10-1. Newton's next letter to Bentley, dated January 17, 1693, contains a mention of this instability:

> *And much harder it is to suppose that all ye particles in an infinite space should be so accurately poised one among another as to stand still in a perfect equilibrium. For I reccon this as hard as to make not one needle only but an infinite number of them (so many as there are particles in an infinite space) stand accurately poised upon their points.*

The Einstein Universe

In 1917, barely two years after formulating his theory of gravity (Chapter 6), Albert Einstein applied his theory toward the ambitious task of constructing a model of the Universe. Like Newton, Einstein also thought that the Universe is static on the large scale. Like Newton's attempted model of the Universe, Einstein's Universe was also imagined to be a homogeneous and isotropic distribution of matter.

The adjectives *homogeneous* and *isotropic* imply the following characteristics. Suppose that, viewed on the large scale, the Universe looks the same from all vantage points. There is *no* preferred observing position in the Universe; all positions are alike. This is the property of *homogeneity*. Furthermore, as we observe the Universe from any such vantage point, should we notice any differences in the structure of the Universe as we look in different directions? If we do not notice any directional differences, then we say that the Universe is *isotropic*. In other words, if you are taken blindfolded from one spot to another in a homogeneous and isotropic universe, after removing your blindfold you cannot tell where you are or in what direction you are looking.

Notice how radically different this concept of the Universe is from the concepts in Greek cosmology. The Greeks attached special significance to the position of the Earth, and from the Earth observers had a unique perspective when looking north, south, east, or west. The Greek universe was neither isotropic nor homogeneous.

Even with these simplifying assumptions about the large-scale structure of the Universe, the *quantitative* details were still lacking in Einstein's model. To determine these details, Einstein needed his theory of gravitation—the general theory of relativity.

In Chapter 5, we saw how the geometry of spacetime is different from Euclid's geometry in the neighborhood of a massive object like the Sun. It is the main feature of general relativity that any distribution of matter (and energy) should affect the geometry of spacetime around it. Einstein therefore expected that the distribution of matter (in the form of stars, galaxies, etc.) should determine the geometry of the large-scale structure of the Universe. But here he encountered one major difficulty.

The equations of general relativity, as obtained by Einstein in 1915, permitted models of the Universe that were homogeneous and isotropic but *not static*. This difficulty is in fact no different from that which troubled Newton two centuries earlier: How can matter remain stationary in spite of its self-gravity? The quote at the beginning of this chapter expresses Newton's difficulty within the framework of his theory of gravity.

To counter the self-gravity of the Universe, Einstein invented a new force of *repulsion*, known as the λ-force. According to Einstein, this force of repulsion increases in direct proportion to the distance between any two chunks of matter. The universal constant λ determines the strength of this force of repulsion.

According to Einstein, the matter in the Universe is held in equilibrium under two opposing forces—the force of attraction of gravity and the λ-force of repulsion. Einstein found that such a Universe could be static, provided it was finite but unbounded.

Can an object be finite but without boundaries? For example, the surface of a sphere has a finite area, but where is its boundary?

An even simpler example is that of a circle: Its circumference is finite, but it has no beginning or end! The circle and the sphere are examples of one- and two-dimensional spaces. Can we similarly imagine a space of *three* dimensions that has a finite *volume* but no boundaries? Although our sense of perception does not help us to form an intuitive picture, we can rely on mathematics. A mathematician would tell us that the space we are trying to imagine is the three-dimensional boundary of a four-dimensional *hypersphere*. Just as the circle and the sphere are made of points at the same distance from a fixed point in space known as the center, so is the hypersphere made of a three-dimensional distribution of points, all of which are located at the same distance from the center. If this distance, known as the radius of the hypersphere, is R, the three-dimensional space of the hypersphere has a volume

$$V = 2\pi^2 R^3$$

This space has no boundary! Just as moving on the surface of a sphere—say, along a meridian (or any great circle)—we eventually come back to our starting place, so would we return to our starting point if we go "straight" in the Einstein Universe (see Figure 10-2). How far would we have to move to make one circuit of the Einstein Universe? The answer is, the distance $2\pi R$.

Einstein's equations determine R in terms of λ. The relation is a simple one:

$$R = \frac{c}{\sqrt{3\lambda}}$$

If we know λ, we know R. How do we know λ?

Einstein's equations give another relation—between λ and the average density of matter, ρ, in the Universe. The relation is

$$\lambda = \frac{4\pi G\rho}{3}$$

So if we know ρ, we know λ, and then we know R. Astronomers tell us that the average density of *visible* matter in the Universe is about 10^{-30} gram per cubic centimeter. This is the density of matter in the form of galaxies, quasars, radio sources, etc.—objects whose existence is established by direct observations. If we also include matter in the form of dark stars and black holes, this estimate may

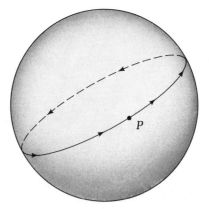

10-2. This figure shows how a light ray in the Einstein Universe would make
a complete round of the Universe and return to the starting point.
A great circle represents a "straight line" in the non-Euclidean
geometry on the surface of a sphere. A light ray constrained to move
in the *two*-dimensional space of the surface of a sphere would travel
along such a track. Thus, it would start from P and end at P.
We have to imagine a similar situation in the *three*-dimensional
space of the Einstein Universe.

go up. But astronomers do not have *any* estimate of what proportion of matter in the Universe is invisible.

Taking $\rho = 10^{-30}$ cm^3, we get $R \cong 3.3 \times 10^{28}$ cm. Thus, one round
of the Einstein Universe is approximately 200 billion light-years
long. That is, light would take 200 billion years to make one round
of the Universe! Einstein's original estimate, based on the information then available, was as low as about 10 million years.

Einstein's Universe was the beginning of modern cosmology.
It pointed the way to further discussion of the large-scale structure
of the Universe *within the framework of physics*.

The Universe is not static

The Einstein Universe was, however, not destined to remain for
long an acceptable model of the actual Universe. In fact, within
twelve years of its birth, astronomers began to discover evidence
that contradicted its basic premise that the Universe is static.

In 1929, Edwin Hubble, an astronomer at the Mount Wilson
Observatory (now part of the Mount Wilson and Las Campanas

10-3. The 100-inch telescope at Mount Wilson, used by Hubble for his investigations of galaxies. (Courtesy of Mount Wilson and Las Campanas Observatories, Carnegie Institution of Washington.)

Observatories) near Pasadena, California, published a paper entitled "A Relation Between Distance and Radial Velocity Among Extragalactic Nebulae" in *The Proceedings of the National Academy of Science of the United States*. This paper presented a remarkable result that was discovered after several years of observations with the newly established 100-inch telescope at Mount Wilson (see Figure 10-3). The observations were based on a systematic study of the spectra of light from galaxies situated well beyond our own Galaxy.*

---

*For a discussion of galaxies, see William J. Kaufmann III, *Galaxies and Quasars* (San Francisco: W. H. Freeman and Company, 1979).

These spectra revealed the redshift effect that we encountered earlier in a different context (Chapter 8). If we look for the familiar lines in the spectrum of a distant galaxy, we find them not at the wavelengths normally associated with these lines in a terrestrial laboratory but at longer wavelengths. For example, the H and K lines of calcium are expected to have wavelengths at 3933 Å and 3968 Å, respectively. (Å stands for the wavelength measure Angstrom, which is a hundred-millionth of a centimeter.) In the spectrum of the galaxy in the Hydra cluster, Hubble and his colleague Milton Humason found these lines at wavelengths 4537 Å and 4578 Å. Thus, by our earlier definition of redshift (see Chapter 8) as the fractional increase in the wavelength, this galaxy has the redshift $z = 0.15$.

Hubble also found another remarkable property of these galaxies. He found that the fainter the galaxy, the larger is its redshift. Now, if we make the assumption (as Hubble did) that the galaxies around us have more or less the same luminosity, then the faintness of a galaxy is an indicator of its distance. The farther the galaxy is from us, the fainter it would look to us.

When Hubble took account of this relationship between faintness and distance, he was able to arrive at a rough method of measuring the distance of nearby galaxies.* He then plotted the redshift of a typical galaxy against its distance from us. Figure 10-4 shows Hubble's graph of redshift z against distance D. Hubble's observational points lay close to the straight line shown in the figure, a result that led him to predict a simple law relating z to D:

$$cz = HD$$

This law, now known as Hubble's law, can be interpreted as follows: If we assume that the redshift of a galaxy is due to the Doppler effect (as Hubble did), then cz gives a measure of the speed of recession of the galaxy. Hubble's law then tells us that the galaxy is moving away from us with a speed that increases in proportion to its distance from us. The constant H, called Hubble's constant, tells us how fast a galaxy is moving at a given distance.

---

*For details of how astronomical distances are measured, see F. Hoyle and J. V. Narlikar, *The Physics–Astronomy Frontier* (San Francisco: W. H. Freeman and Company, 1980).

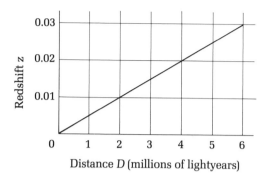

10-4. The straight line graph calculated by Hubble from his original measurements of the redshift z and the distance D of galaxies. Later it
turned out that Hubble had grossly underestimated the distances of
galaxies. The present estimates (which are still subject to uncertainties)
make these galaxies five to ten times *farther away* than Hubble's
original estimates.

Thus, Hubble's law presents a picture of the Universe far different from the *static* Universe of Einstein. The Universe seems to
be *dynamic*, with galaxies moving away from us, as if we are in
a highly unpopular part of the Universe. Does this mean that we
are back to some revised version of the old Greek cosmology that
accorded a unique status to us on the Earth as observers of the
cosmos?

The Big Bang

Hubble's law does not, however, take us back to pre-Copernican
days. Although the redshift of a galaxy appears to increase with its
distance from us, this effect does not necessarily place us in any
unique position. In fact, if we imagine ourselves to be observing
the Universe from another galaxy, we would find that the same
recession phenomenon holds true with respect to our new vantage
point. In other words, all galaxies would serve equally well as
observation posts for the Hubble effect—our Galaxy does not enjoy
any special status.

A common way of describing the recession of galaxies is to say
that *the Universe is expanding*. The space in which the galaxies are

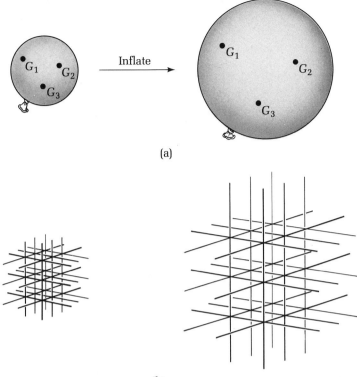

(a)

(b)

10-5. Two examples of expansion analogous to the expanding Universe.
(a) As we inflate a balloon, the dots G_1, G_2, and G_3 on its surface move
away from each other. No particular dot can claim a unique position
on the surface of the expanding sphere. (b) A grid of metal wires
expands when heated. The lattice points all move away from each other.
Again, no one point can claim a privileged position within the grid.
(From *The Physics–Astronomy Frontier* by F. Hoyle and J. V. Narlikar.
Copyright © 1980. W. H. Freeman and Company.)

embedded is expanding, so that the separation between any two
galaxies is increasing. Figure 10-5 illustrates this expansion effect
with two examples: (a) If we blow up a spotted balloon, the spots
appear to move away from one another; yet there is no particular
spot that can claim a special status. (b) A cubical grid of metal
wires expands when heated, so that the lattice points of the grid
move farther apart from one another. A third example is a special
kind of plastic strip that expands to three times its size when
heated in an oven. Any figure drawn on the strip will also expand.

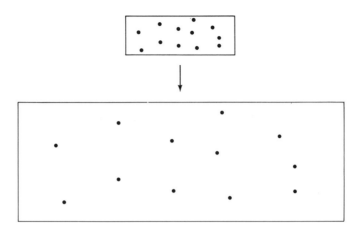

10-6. The dots on the plastic strip move apart from one another as the strip expands to three times its size (both in length and width).

In Figure 10-6, we see such a strip with a few dots on it, before and after heating. The dots have moved away from one another as the galaxies do in the expanding Universe.

Once we accept the fact that the galaxies are receding from each other, the physicist will naturally ask what the cause of this expansion is. Has the expansion been going on in the past? Will it continue in the future? Questions like these can be answered with the help of mathematical models of the expanding Universe, just as Einstein constructed a model of the static Universe. Here again, Einstein's general relativity comes to the help of the theoretician.

Even before Hubble published his results, the Russian physicist Alexander Friedmann had constructed such models. Friedmann used Einstein's assumptions about the homogeneity and isotropy of the Universe, but he dropped the assumption of a *static* Universe. The Friedmann models do not require the λ-term introduced by Einstein for his static model; they are based on Einstein's old equations of 1915.

If gravity is the only force involved in these cosmological models, and if gravity has a tendency to contract an object rather than to expand it, how was Friedmann able to construct models of an *expanding* Universe? The answer to this question can be understood with the help of a simple example. Recall our ball thrower in Chapter 3. The ball thrower sends a ball *upward* in the vertical direction. For a time the ball travels upward, although the force of Earth's gravity tends to pull it *down*. The reason the ball can go up at all is because it has been thrown with an *initial* upward velocity. It will continue to go up until its store of kinetic energy is completely exhausted.

In the same way, the Universe tends to expand in spite of gravity because in the *initial* stages it was given large kinetic energy in an outward explosion. Imagine a gigantic cosmic explosion in which the components of the Universe were thrown apart. What we see today is the debris of this cosmic explosion, commonly known as the *Big Bang*. The Friedmann Universe was *created* with a bang.

Had there been no gravity, the speeds of recession of the galaxies would have continued unabated. However, the gravity of the Universe shows its effect by *slowing down* these recession speeds, just as the Earth's gravity slows down the ball's upward speed. We can follow this ball example further to determine the future behavior of the Universe.

We saw in Chapter 3 that the ball falls back to Earth unless it is thrown with sufficient speed. The critical speed that determines whether the ball will fall back or not is the *escape speed*. If the escape speed is exceeded, the ball never returns to Earth; it keeps going away from Earth forever.

Similarly, there is a critical feature in the Big Bang that determines whether the Universe will keep on expanding forever, albeit with decreasing speed, or whether it will slow to a halt and fall back upon itself. As expected from the geometrical bias of Einstein's theory, this dynamical behavior of the Universe turns out to be related to its geometry.

Is the Universe open or closed?

The space of the static Einstein Universe is finite in volume but unbounded. A ray of light going in any direction in this Universe makes a complete round of space and returns to the starting point. A Universe with these properties is said to be *closed*. The curvature of space in such a Universe is positive (see Chapter 5).

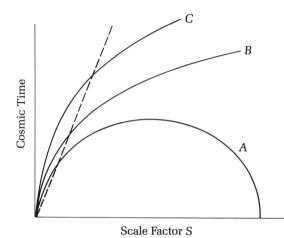

Cosmic Time

Scale Factor S

10-7. The three types of cosmological models. Model *A* describes a closed Universe, whose expansion slows and gives way to contraction. *B* and *C* are models of an open Universe, which goes on expanding forever. The dashed lines represent the expansion of the Universe in the absence of gravity.

The space in the expanding Friedmann Universe can have positive, zero, or negative curvature. We denote these three possibilities by *A*, *B*, and *C*, respectively. The Friedmann model Universe of type *A* is closed, as is the Einstein Universe. The models of types *B* and *C* are *open*. They are *infinite and unbounded*. All three types, however, satisfy the condition of homogeneity and isotropy at any epoch. That is, at any epoch we can locate observers all over the Universe who will report the same large-scale view of the cosmos. A typical observer will also find that the Universe looks the same in all directions.*

How the differences of geometry affect the expansion of the Universe is shown in Figure 10-7, where *S* is the characteristic

---

*These large-scale views describe the structure of the Universe on the scale of, say, a few million light-years or so. On a smaller scale, we do notice inhomogeneities, such as the shape of our Galaxy, our own off-center location in it, etc. These "local irregularities" are irrelevant to discussions of the dynamics of the Universe on the large scale.

scale factor of the expanding Universe. The relative linear separation between two typical galaxies changes in proportion to S, so that an increase of S with time signifies the expansion of the Universe.

In Figure 10-7, all three curves for models A, B, and C start off with S = 0 and have S subsequently increasing with time. In the case of A, however, the expansion slows to a halt and gives way to contraction. Thus, the closed model of the Universe has a *contraction phase* following the expansion phase. The open models continue to expand forever. Compare this behavior of the Universe with the flight of the ball we discussed earlier.

All three curves show the expected slowing of expansion due to gravity. The epoch at which S was zero is known as the *Big Bang Epoch*, the instant of creation when all the matter in the Universe was compressed in zero volume. This is the singular instant that marks the origin of the Universe. We may start the cosmic clock at this instant, so that in Figure 10-7 the time axis reads t = 0 at S = 0.

The singular epoch at t = 0 and the subsequent expansion of the Universe bear a great similarity to the white hole theory of Chapter 9. At the singularity, any finite chunk of the Universe is like a white hole erupting from a point in space. This similarity led to the suggestion that white holes may be "delayed bangs"—eruptions that took place in space later than the moment t = 0.

The dashed straight line of Figure 10-7 shows how S would have behaved had there been no gravity. In fact, there is an empty Friedmann model that shows this rate of expansion. The greater the density of matter in the Universe, the greater the effect of gravity on slowing the expansion. We therefore expect the closed models to be *denser* than the open models. There is a critical density that decides whether the Universe is open or closed. At the present epoch, this density, sometimes known as the *closure density*, is given by

$$\rho_c = \frac{3H^2}{8\pi G}$$

If the actual density in the Universe *exceeds* ρ_c, the Universe is closed; if it equals ρ_c or is less than ρ_c, the Universe is open. The three models A, B, and C of Figure 10-7 have $\rho > \rho_c$, $\rho = \rho_c$, and $\rho < \rho_c$.

The question of whether the Universe is open or closed can in principle be decided if we have reasonably accurate measures of the density ρ of matter in the Universe and the Hubble constant, H. Unfortunately, neither ρ nor H are known accurately. We earlier gave a value of $\rho \cong 10^{-30}$ cm^3 for the visible matter in the Universe. What is the value of H?

Scientists differ in their answers to this question. The quantity $1/H$ has the dimension of time, and it represents the upper limit on the age of the Universe. That is, since the origin in a Big Bang at $t = 0$, the time elapsed to the present epoch cannot exceed the value $1/H$. On the basis of the present data, the range within which $1/H$ is expected to lie extends from 10 billion to 20 billion years.

Taking the lower end of this range, we can calculate the density $\rho_c \approx 10^{-29}$ cm^3. If we had taken the upper end of the range, we would have arrived at four times this value of ρ_c. Notice that in either case $\rho < \rho_c$, and so we have a prima facie case for the open Universe of type C.

However, the issue of whether the Universe is open or closed is not as simple as that! The proponents of the closed Universe argue that the estimate of ρ given above is only a lower estimate. If there is considerably more hidden matter in the Universe, the estimate for ρ may be higher than the above value. If it becomes high enough to exceed ρ_c, the Universe may be closed.

Another way to decide which of the models A, B, or C is the correct one is to measure the rate at which the expansion of the Universe is slowing down. The rate of slowing down is higher for A-type models and lower for C-type models than it is for B-type models. However, this is a difficult measurement to make, and it has not yet been possible to arrive at unambiguous conclusions.

Was there a Big Bang?

Cosmologists are often asked whether there is any direct astronomical evidence for the Big Bang.

When astronomers survey the distant parts of the Universe, they do not see the Universe as it is *now* but as it was in the remote past. This happens because astronomical observations depend on light, which has a finite velocity. Light that carries information from a

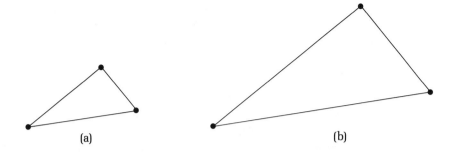

10-8. In (a) we see a triangle formed by three galaxies. In (b) the triangle has expanded to twice its original size, so that each galaxy is twice as far away from the other two galaxies. This is the result of the expansion of the Universe. If light from any of these galaxies left at the epoch of (a) and was received at the epoch of (b), this light would show a redshift z = 1. (This result is a special case of the general result described in the text.)

distant source takes some time to get to the astronomer. For example, light from a galaxy a billion light-years away from here takes a billion years to reach here. Consequently, the astronomer sees the galaxy as it was a billion years ago.

This delay due to the finite speed of light is hardly noticeable in our day-to-day observations on the Earth because the distances are small. However, people making international telephone calls routed via satellite do notice a momentary delay in conversational responses, which are carried by radio waves with the speed of light.

Because the Universe has been expanding since its origin at $t = 0$, the scale factor S was smaller in the past than it is now. Is there any direct way we can compare the scale factor a billion years ago with the scale factor at present? Yes! A simple result from Einstein's theory of general relativity enables us to do this. Suppose we measure the redshift of a galaxy at the distance of a billion light-years and call it z. Then the ratio of the present scale factor to the scale factor a billion years ago is just the quantity $1 + z$ (see Figure 10-8).

Thus, the larger the redshift of the observed object, the smaller the scale factor of the Universe when the light ray left that object.

The Big Bang epoch therefore corresponds to the epoch of infinite redshift.

The largest redshift measured to date is $z = 3.53$, of quasar OQ 172. There is currently controversy as to how far away quasars are. But even assuming that they are as distant as they appear to be, the farthest known object does not take us all that far back into the past!

There is, however, indirect evidence that takes us much farther back in the history of the Universe. We end this chapter with a brief discussion of this important evidence.

We have so far concerned ourselves with only the geometrical aspects of the Universe. Cosmology goes beyond this; it is also concerned with the physical aspects of the Universe. In the 1940s, George Gamow first discussed the physical state of the Universe just a few moments after the Big Bang.

The Big Bang implies violent activity, and physicists have wondered what form this activity took. Such speculations take us to the most elementary states of matter—perhaps even to *quarks*, which may have existed before electrons, protons, and other elementary particles were formed. Gamow's discussion, however, began at the stage after these particles were formed. In the early stages, elementary particles existed in a heat bath of extremely high temperature, of the order of 10 billion degrees Kelvin, just 1 second after the Big Bang. This early stage had another remarkable property—it featured the dominance of radiation over matter, in contrast to the present state of the Universe, which is dominated by matter. Figure 10-9 shows how the transition from the early radiation-dominated era to the later matter-dominated era occurred in the course of expansion of the Universe.

Gamow conjectured that the relic of the early hot radiation should now be seen in the form of cool radiation. How cool? On the basis of the information then available, Gamow and his colleagues Ralph Alpher and Robert Hermann predicted the present temperature of the radiation to be about 5 degrees Kelvin. At this temperature, the radiation would exist predominantly in the form of *microwaves*.

The first important indications of the existence of such radiation came in 1965, when Arno Penzias and Robert Wilson at the Bell Telephone Laboratories in Holmdale, New Jersey, accidentally

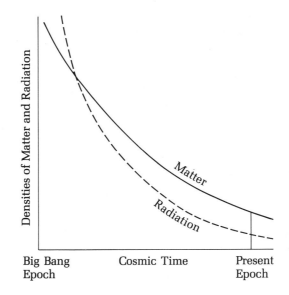

Big Bang Epoch   Cosmic Time   Present Epoch

10-9. A schematic diagram showing how the density of matter (*solid curve*) and radiation (*dashed curve*) decrease as the Universe expands. In the early states, the Universe was dominated by radiation; the present epoch is dominated by matter. This is because as the Universe expands, the decline in the density of radiation is more rapid than the decline in the density of matter. (Figure not drawn to scale.)

found a radiation background at 7 cm wavelength, with a temperature about 3.5 on the absolute scale. The possible cosmological significance of this discovery prompted several groups of astronomers to measure the radiation background at other wavelengths. Today measurements exist at wavelengths ranging from 70 cm on the long-wavelength end to wavelengths shorter than 1 mm. The latter measurements became possible only with recent advances in space technology.

The characteristics of this radiation are those expected for the radiation left over from the Big Bang. The temperature is not too far off from the value roughly predicted by Gamow and his colleagues. The radiation appears as a background—that is, it seems to come equally from all directions in the Universe, thus suggesting its cosmic origin.

At present, there are some matters still to be settled. If the questions regarding the various interpretations are satisfactorily resolved, astronomers can point to the radiation background as indirect evidence for the Big Bang and for the models of the Universe based on Einstein's theory of gravity.

Golconde (1953) by the Belgian artist René Magritte (1898–1967). (Private collection, U.S.A. Photographer: Taylor and Dull, N.Y.)

11

The Many Faces of Gravity

11

The Many Faces of Gravity

Our discussion of gravity began with the falling apple and has taken us from ocean tides to the planets, comets, and satellites of the solar system, to the different stages in the evolution of a star, to the curved spacetime of general relativity, to the weird effects associated with black holes and white holes, and finally to the large-scale structure of the Universe itself. None of the other basic forces of physics has such a wide range of applications. Although gravity is by far the weakest of the four known basic forces, its effects are the most dramatic.

Indeed, it would be an amusing exercise to speculate on the state of the world if there were no gravity at all! Would atoms and molecules be affected? As far as we know, the presence or absence of gravity does not play a crucial role in the existence and stability of the microworld. The strong, weak, and electromagnetic forces are the main forces at this level. Even at the macroscopic level of the objects we see around us in our daily lives, gravity does not appear to play a crucial role. After all, even astronauts have demonstrated that they can live in simulated conditions of weightlessness. Neither the astronauts nor their spacecrafts come apart in such circumstances. The basic binding force at this level is the force of electricity and magnetism.

But we can go no further in dispensing with gravity. If we eliminate gravity on a bigger scale, disasters lie in store. With the

Earth's gravity gone, there is no force to bind the Earth together as a spherical object or to keep us on its surface. The living systems on the Earth have complex biological systems that have evolved with and have adapted themselves to the explicit presence of the Earth's gravity. Without gravity, the Earth would also lose its protective atmosphere.

On a still larger scale, without gravity the Earth would no longer be attracted by the Sun and would take off in a straight path instead of going around the Sun. The Sun itself would no longer be able to maintain stability but would disperse outward. Without gravity, neither the Sun nor any other stars could exist, nor could larger systems such as galaxies and quasars. These are just a few speculations on a zero-gravity Universe. Incomplete as they are, they still give us some indication of the importance of gravity to the physical world.

In spite of its importance, gravity remains shrouded in mystery. Having stated the inverse-square law of gravity, Newton declined to further try to elucidate *why* this law operates. Einstein provided an ingenious connection between gravity and the geometry of spacetime, but even he was conscious of the fact that his description of gravity placed it farther apart from the rest of physics. Einstein's prescription eliminated gravity as a force. Instead of simply affecting the motion or equilibrium of a body as any normal physical force is expected to do, gravity transforms the geometry of spacetime around the body. To bridge the gap between gravity and other physical forces, Einstein hoped to construct a *unified field theory* of all physical interactions. He was unsuccessful in this ambitious task, in spite of several years of research.

Nevertheless, the goal of unification continues to tempt and challenge physicists today. The recent success of the Weinberg–Salam theory (named after Stephen Weinberg and Abdus Salam, who independently proposed the theory) in unifying the electromagnetic force and the weak force has prompted many theoreticians to think that a unified theory for all physical interactions is near. There are encouraging signs that even the strong force that binds particles in the atomic nucleus may be linked with the electromagnetic and weak forces in the future. Physicists call such a unification a *grand unification scheme*.

But gravity still remains far apart from this chain of developments, partly because of its unusual description as a geometrical effect of spacetime rather than as a straightforward force. To some extent, the difficulty also lies in identifying what the quantum effects of gravity are. The quantum theory usually relates to the microscopic world. At the particle level, gravity is very weak. How can physicists study such effects?

As we increase the speed of a particle, its energy also increases. As a rule, this increases the strength of its interaction with other particles. Physicists use particle accelerators to study the nature of various particle interactions by firing particles at each other with high energy. At particle energies of 10^{17} electron volts, the weak force becomes appreciably large—large enough to be comparable to the electromagnetic force. Our present high-energy particle accelerators just manage to accelerate particles to one hundred-thousandth of this energy. By comparison, however, the energy at which gravity will be of the same order of strength as, say, the electromagnetic force is about 10^{46} electron volts, which is some 32 orders of magnitude higher than the energy limit of existing accelerators.

These numbers illustrate the difficulties of making progress in understanding the nature of gravity in the terrestrial laboratory. These difficulties in turn imply that, for a long time to come, further understanding of this mysterious interaction must come through astronomy.

We have discussed in this book a few highlights of the astronomical effects of gravity. I now close this discussion by enumerating some unresolved features of these effects.

Over the past decade, considerable research has been done on the physics and astrophysics of black holes. We discussed some of this work in Chapters 7 through 9. Some enthusiasts believe that black holes are the ultimate solution to the energy problem. However, the skeptics remain unconvinced that black holes even exist or have been detected. Defenders of black holes say that the evidence is necessarily circumstantial, and nothing more can be expected. Nevertheless, the note of caution that should accompany any circumstantial evidence or a proper consideration of alternatives to the black-hole theory is often missing from the deduction that "a black hole exists in such-and-such object."

White holes, by comparison, are theoretically directly observable. Many astronomers have serious doubts about their existence, however, and there is no proper theory to describe under what circumstances a white hole should form.

Finally, it is clear that the Big Bang theory of the Universe leaves the very important question of the origin of the Universe unanswered. Why did the Big Bang occur? Why, how, and when did matter first appear in the Universe? Can we really trust the present laws of physics to such an extent that we can deduce what the Universe was like at the time of the Big Bang?

Many supporters of the Big Bang Universe consider the event of Creation to be beyond the domain of science. This belief is reminiscent of Newton's approach. When faced with difficulties of this kind, he postulated "Divinity" as the solution. Newton's departure from the scientific approach on such occasions pleased the devout but drew protests from his contemporary scientists (such as Leibnitz).

Questions about the origin of the Universe continue to bother some physicists. Because of these questions, the *Steady-State theory* of the Universe was proposed by Hermann Bondi, Thomas Gold, and Fred Hoyle in 1948. This is a Universe without a beginning and without an end. The matter in this Universe is created not all at once in a single explosion, as in the Big Bang Universe, but continuously in small amounts throughout the Universe at all times.

The Steady-State Universe is at present under a cloud, largely because within its framework no satisfactory explanation has been found for the microwave background radiation described in Chapter 10. The discovery of this radiation appeared to favor the Big Bang Universe, yet many questions still remain unresolved.

And here we leave our study of the many faces of gravity. Through the study of gravity, nature has so far revealed many of her secrets, but many more are being reserved for the future.

Index